Encyclopaedia of

Genes, Genomes, Genetics and Chromosomes

Encyclopaedia of

Genes, Genomes, Genetics and Chromosomes

Volume 4

Editor:
Dr. R.N. Shukla

AGROTECH PRESS
JAIPUR ▪ NEW DELHI

AGROTECH PRESS

Regd. Off. "Bony Residency", Gate No. 2
Opp. Tilak Public School, Vishweswariya Nagar,
Gopalpura Road, Jaipur - 302018, India
Ph: 0141 - 2761280, Telefax: 2761381
E-mail : agrotechpress@gmail.com

Delhi Office: 102, 1st Floor, Satyam House,
4327/3, Ansari Road, Darya Ganj,
New Delhi - 110002, India
Ph: 011-45652440
visit us: www.agrotechpress.com

Distributed By
KEYA INTERNATIONAL PRESS LLP
E-mail: keyainternationalpressllp@gmail.com

Published in India, 2024

ISBN : 9789384568238 (5 Vols. Set)

Typeset by Thomson Press

Printed and bound in India by Thomson Press Ltd.,
New Delhi

Contents

Preface

Genetic engineering refers to a set of technologies that are being used to change the genetic makeup of cells and move genes across species boundaries to produce novel organisms. The techniques involve highly sophisticated manipulations of genetic material and other biologically important chemicals.

Genes are the chemical blueprints that determine an organism's traits. Moving genes from one organism to another transfers those traits. Through genetic engineering, organisms are given new combinations of genes—and therefore new combinations of traits—that do not occur in nature and, indeed, cannot be developed by natural means. Such an artificial technology is radically different from traditional plant and animal breeding. Nature can produce organisms with new gene combinations through sexual reproduction. A brown cow bred to a yellow cow may produce a calf of a completely new colour. But reproductive mechanisms limit the number of new combinations. Cows must breed with other cows. A breeder who wants a purple cow would be able to breed towards one only if the necessary purple genes were available somewhere in a cow or a near relative to cows. A genetic engineer has no such restriction. If purple genes are available anywhere in nature—in a sea urchin or an iris—those genes could be used in attempts to produce purple cows. This unprecedented ability to shuffle genes means that genetic engineers can concoct gene combinations that would never be found in nature. Genetic engineering has been said to be no different than ancient breeding methods but this is untrue. For a start, breeding or cross-breeding, or in-breeding all work by using the same species. In contrast genetic engineering allows us to combine fish, mouse, human and insect genes in the same person or animal. Genetic engineering therefore has few limits - except our imagination, and our moral or ethical code. Genetic engineering makes the whole digital revolution look nothing. Digital technology changes what we do. Genetic engineering has the power to change who we are.

Human cloning is a type of genetic engineering, but is not the same as true genetic manipulation. In human cloning, the aim is to duplicate the genes of an existing person so that an identical set is inside a human egg. The result is intended to be a cloned twin, perhaps of a dead child. Genetic engineering in its fullest form would result in the child produced having unique genes - as a result of laboratory interference, and therefore the child will not be an identikit twin. Genetic engineering could create crops that grow in desert heat, or without fertiliser. Genetic engineering could make bananas or other fruit which contain vaccines or other medical products. In medicine genetic engineering has been used to mass-produce insulin, human growth hormones, follistim, human albumin, monoclonal antibodies, antihemophilic factors, vaccines and many other drugs. Vaccination generally involves injecting weak live, killed or inactivated forms of viruses or their toxins into the person being immunized. Genetically engineered viruses are being developed that can still confer immunity, but lack the infectious sequences. Mouse hybridomas, cells fused together to create monoclonal antibodies, have been humanised through genetic engineering to create human monoclonal antibodies. Genetic engineering is used to create animal models of human diseases. Genetically modified mice are the most common genetically engineered animal model. They have been used to study and model cancer, obesity, heart disease, diabetes, arthritis, substance abuse, anxiety, aging and Parkinson disease. Potential cures can be tested against these mouse models. Also genetically modified pigs have been bred with the aim of increasing the success of pig to human organ transplantation.

This book presents the principles of genetic engineering and its associated techniques in sufficient details. Besides useful as a textbook to undergraduate and postgraduate students, it is useful to many experienced researchers as an informative reference source.

—Editor

1

Scattering Techniques

Elastic Scattering

Materials that do not have long range order may also be studied by scattering methods that rely on elastic scattering of monochromatic X-rays.

- Small angle X-ray scattering (SAXS) probes structure in the nanometer to micrometer range by measuring scattering intensity at scattering angles 2è close to 0°.
- X-ray reflectivity is an analytical technique for determining thickness, roughness, and density of single layer and multilayer thin films.
- Wide angle X-ray scattering (WAXS), a technique concentrating on scattering angles 2è larger than 5°.

Inelastic Scattering

When the energy and angle of the inelastically scattered X-rays are monitored scattering techniques can be used to probe the electronic band structure of materials.

- Compton scattering
- Resonant inelastic X-ray scattering (RIXS)
- X-ray Raman scattering
- Nucleic acid double helix

In molecular biology, the term double helix refers to the structure formed by double-stranded molecules of nucleic acids such as DNA and RNA. The double helical structure of a nucleic acid complex arises as a consequence of its secondary structure, and is a fundamental component in determining its tertiary structure. The term entered popular culture with the publication in 1968 of The Double Helix: A Personal Account of the Discovery of the Structure of DNA, by James Watson. The DNA double helix is a spiral polymer of nucleic acids, held together by nucleotides which base pair together. In B-DNA, the most common double helical structure, the double helix is right-handed with about 10–10.5 nucleotides per turn. The double helix structure of DNA contains a major groove and minor groove, the major groove being wider than the minor groove. Given the difference in widths of the major groove and minor groove, many proteins which bind to DNA do so through the wider major groove.

History

The double-helix model of DNA structure was first published in the journal Nature by James D. Watson and Francis Crick in 1953, (X,Y,Z coordinates in 1954) based upon the crucial X-ray diffraction image of DNA labeled as "Photo 51", from Rosalind Franklin in 1952, followed by her more clarified DNA image with Raymond Gosling, Maurice Wilkins, Alexander Stokes, and Herbert Wilson, as well as base-pairing chemical and biochemical information by Erwin Chargaff. The previous model was triple-stranded DNA. The realization that the structure of DNA is that of a double-helix elucidated the mechanism of base pairing by which genetic information is stored and copied in living organisms, and is widely considered one of the most important scientific discoveries of the 20th century. Crick, Wilkins, and Watson each received one third of the 1962 Nobel Prize in Physiology or Medicine for their contributions to the discovery. (Franklin, whose breakthrough X-ray diffraction data was used to formulate the DNA structure, died in 1958, and thus was ineligible to be nominated for a Nobel Prize.)

Nucleic Acid Thermodynamics

Nucleic acid thermodynamics is the study of the thermodynamics of nucleic acid molecules, or how temperature affects nucleic acid structure. For multiple copies of DNA molecules, the melting temperature (T_m) is defined as the temperature at which half of the DNA strands are in the double-helical state and half are in the random coil states. The melting temperature depends on both the length of the molecule, and the specific nucleotide sequence composition of that molecule.

Concepts

Hybridization

Hybridization is the process of establishing a non-covalent, sequence-specific interaction between two or more complementary strands of nucleic acids into a single hybrid, which in the case of two strands is referred to as a duplex. Oligonucleotides, DNA, or RNA will bind to their complement under normal conditions, so two perfectly complementary strands will bind to each other readily. In order to reduce the diversity and obtain the most energetically preferred hybrids, a technique called annealing is used in laboratory practice. However, due to the different molecular geometries of the nucleotides, a single inconsistency between the two strands will make binding between them less energetically favorable.

Measuring the effects of base incompatibility by quantifying the rate at which two strands anneal can provide information as to the similarity in base sequence between the two strands being annealed. The hybrids may be dissociated by thermal denaturation, also referred to as melting. Here, the solution of hybrids is heated to break the hydrogen bonds between nucleic bases, after which the two strands separate. In the absence of external negative factors, the processes of hybridization and melting may be repeated in succession indefinitely, which lays the ground for polymerase chain reaction. Most commonly, the pairs of nucleic bases A=T and Ga"C are formed, of which the latter is more stable.

Denaturation

DNA denaturation, also called DNA melting, is the process by which double-stranded deoxyribonucleic acid unwinds and separates into single-stranded strands through the breaking of hydrogen bonding between the bases. Both terms are used to refer to the process as it occurs when a mixture is heated, although "denaturation" can also refer to the separation of DNA strands induced by chemicals like urea. The process of DNA denaturation can be used to analyze some aspects of DNA. Because cytosine / guanine base-pairing is generally stronger than adenosine / thymine base-pairing, the amount of cytosine and guanine in a genome (called the "GC content") can be estimated by measuring the temperature at which the genomic DNA melts. Higher temperatures are associated with high GC content.

DNA denaturation can also be used to detect sequence differences between two different DNA sequences. DNA is heated and denatured into single-stranded state, and the mixture is cooled to allow strands to rehybridize. Hybrid molecules are formed between similar sequences and any differences between those sequences will result in a disruption of the base-pairing. On a genomic scale, the method has been used by researchers to estimate the genetic distance between two species, a process known as DNA-DNA hybridization. In the context of a single isolated region of DNA, denaturing gradient gels and temperature gradient gels can be used to detect the presence of small mismatches between two sequences, a process known as temperature gradient gel electrophoresis. Methods of DNA analysis based on melting temperature have the disadvantage of being proxies for studying the underlying sequence; DNA sequencing is generally considered a more accurate method.

The process of DNA melting is also used in molecular biology techniques, notably in the polymerase chain reaction (PCR). Although the temperature of DNA melting is not diagnostic in the technique, methods for estimating T_m are important for determining the appropriate temperatures to use in a protocol.

DNA melting temperatures can also be used as a proxy for equalizing the hybridi-zation strengths of a set of molecules, e.g. the oligonucleotide probes of DNA microarrays.

Annealing

Annealing, in genetics, means for DNA or RNA to pair by hydrogen bonds to a complementary sequence, forming a double-stranded polynucleotide. The term is often used to describe the binding of a DNA probe, or the binding of a primer to a DNA strand during a polymerase chain reaction (PCR). The term is also often used to describe the reformation (renaturation) of complementary strands that were separated by heat (thermally denatured). Proteins such as RAD52 can help DNA anneal.

Methods for Estimating Melting Temperatures

Several formulas are used to calculate T_m values. Some formulas are more accurate in predicting melting temperatures of DNA duplexes. One problem in nucleic acid thermodynamics is to determine the thermodynamic parameters for forming double-stranded nucleic acid AB from single-stranded nucleic acids A and B.

$$AB \leftrightarrow A + B$$

The equilibrium constant for this reaction is $K = \frac{[A][B]}{[AB]}$. According to thermodynamics, the relation between free energy, ΔG, and K is $\Delta G° = -RT\ln K$, where R is the ideal gas law constant, and T is the kelvin temperature of the reaction. This gives, for the nucleic acid system,

$$\Delta G^o = -RT\ In \frac{[A][B]}{[AB]}.$$

The melting temperature, T_m, occurs when half of the double-stranded nucleic acid has dissociated. If no additional nucleic acids are present, then [A], [B], and [AB] will be equal, and equal to half the initial concentration of double-stranded nucleic acid, $[AB]_{initial}$. This gives an expression for the melting point of a nucleic acid duplex of

$$T_m = \frac{-\Delta G^o}{R\ In \frac{[AB]_{initial}}{2}}.$$

Because $\Delta G° = \Delta H° - T \Delta S°$, T_m is also given by

$$T_m = \frac{\Delta H^o}{\Delta S^o - R\ In\ \frac{[AB]_{intial}}{2}}.$$

The terms $\Delta H°$ and $\Delta S°$ are usually given for the association and not the dissociation reaction (see the nearest-neighbor method for example). This formula then turns into :

$$T_m = \frac{\Delta H^o}{\Delta S^o + R\ In\left([A]_{total} - [B]_{total}/2\right),} where [B]_{total} < [A]_{total}.$$

This equation is based on the assumption that only two states are involved in melting: the double stranded state and the random-coil state. However, nucleic acids may melt several intermediate states. To account for such complicated behavior, the methods of statistical mechanics must be used.

Nucleic Acid Hybridization

Hybridization is the process of complementary base pairs binding to form a double helix. Melting is the process by which the interactions between the strands of the double helix are broken, separating the two nucleic acid strands. These bonds are weak, easily separated by gentle heating, enzymes, or physical force. Melting occurs preferentially at certain points in the nucleic acid. T and A rich sequences are more easily melted than C and G rich regions. Particular base steps are also susceptible to DNA melting, particularly T A and T G base steps. These mechanical features are reflected by the use of sequences such as TATAA at the start of many genes to assist RNA polymerase in melting the DNA for transcription.

Strand separation by gentle heating, as used in PCR, is simple providing the molecules have fewer than about 10,000 base pairs (10 kilobase pairs, or 10 kbp). The intertwining of the DNA strands makes long segments difficult to separate. The cell avoids this problem by allowing its DNA-melting enzymes (helicases) to work concurrently with topoisomerases, which can chemically cleave the phosphate backbone of one of the strands so that it can swivel around the other. Helicases unwind the strands to facilitate the advance of sequence-reading enzymes such as DNA polymerase.

Base Pair Geometry

The geometry of a base, or base pair step can be characterized by 6 coordinates: Shift, slide, rise, tilt, roll, and twist. These values precisely define the location and orientation in space of every base or base pair in a nucleic acid molecule relative to its predecessor along the axis of the helix. Together, they characterize the helical structure of the molecule. In regions of DNA or RNA where the "normal" structure is disrupted, the change in these values can be used to describe such disruption.

For each base pair, considered relative to its predecessor, there are the following base pair geometries to consider:

- Shear
- Stretch
- Stagger
- Buckle
- Propeller twist: rotation of one base with respect to the other in the same base pair.
- Opening
- Shift: displacement along an axis in the base-pair plane perpendicular to the first, directed from the minor to the major groove.
- Slide: displacement along an axis in the plane of the base pair directed from one strand to the other.
- Rise: displacement along the helix axis.
- Tilt: rotation around this axis.
- vRoll: rotation around this axis.
- Twist: rotation around the helix axis.
- vx-displacement
- y-displacement
- inclination
- tip
- pitch: the number of base pairs per complete turn of the helix.

Rise and twist determine the handedness and pitch of the helix. The other coordinates, by contrast, can be zero. Slide and shift are typically small in B-DNA, but are substantial in A- and Z-DNA. Roll and tilt make successive base pairs less parallel, and are typically small. A diagram of these coordinates can be found in 3DNA website.

Note that "tilt" has often been used differently in the scientific literature, referring to the deviation of the first, inter-strand base-pair axis from perpendicularity to the helix axis. This corresponds to slide between a succession of base pairs, and in helix-based coordinates is properly termed "inclination".

Helix Geometries

At least three DNA conformations are believed to be found in nature, A-DNA, B-DNA, and Z-DNA. The "B" form described by James D. Watson and Francis Crick is believed to predominate in cells. It is 23.7 Å wide and extends

34 Å per 10 bp of sequence. The double helix makes one complete turn about its axis every 10.4-10.5 base pairs in solution. This frequency of twist (known as the helical pitch) depends largely on stacking forces that each base exerts on its neighbours in the chain.

A-DNA and Z-DNA differ significantly in their geometry and dimensions to B-DNA, although still form helical structures. The A form appears likely to occur only in dehydrated samples of DNA, such as those used in crystallographic experiments, and possibly in hybrid pairings of DNA and RNA strands.

Segments of DNA that cells have been methylated for regulatory purposes may adopt the Z geometry, in which the strands turn about the helical axis the opposite way to A-DNA and B-DNA. There is also evidence of protein-DNA complexes forming Z-DNA structures.

Other conformations are possible; A-DNA, B-DNA, C-DNA, E-DNA, L-DNA (the enantiomeric form of D-DNA), P-DNA, S-DNA, Z-DNA, etc. have been described so far. In fact, only the letters F, Q, U, V, and Y are now available to describe any new DNA structure that may appear in the future. However, most of these forms have been created synthetically and have not been observed in naturally occurring biological systems. There are also triple-stranded DNA forms and quadruplex forms such as the G-quadruplex.

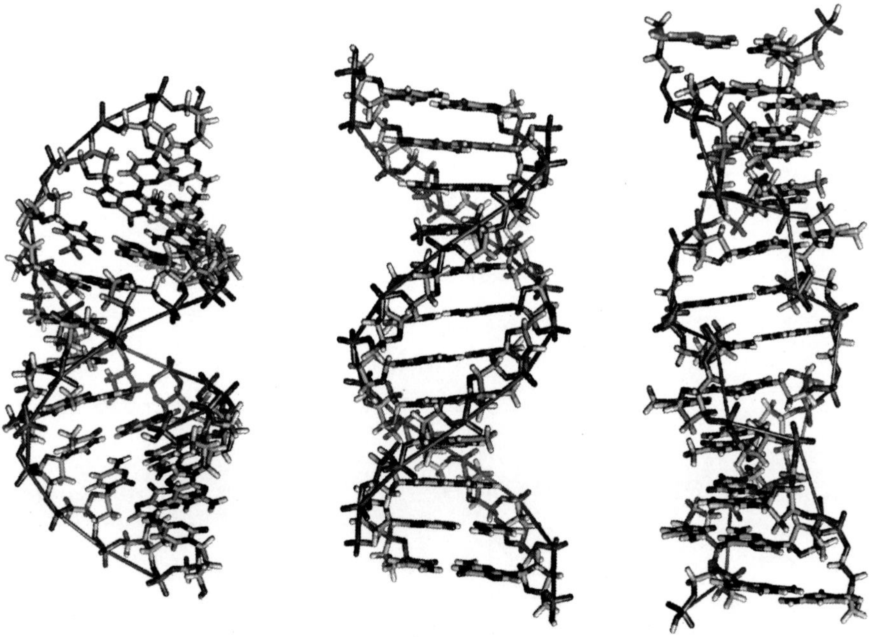

Figure: *The structures of A-, B-, and Z-DNA.*

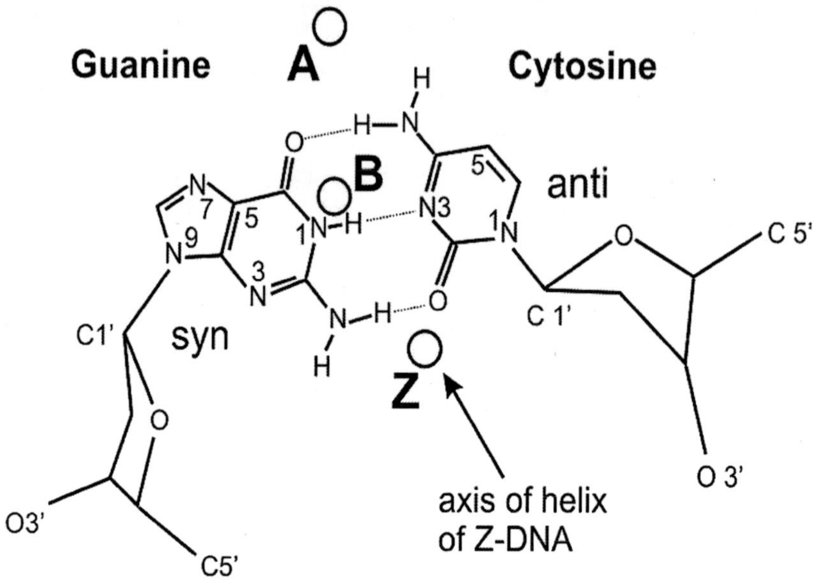

Figure: *The helix axis of A-, B-, and Z-DNA.*

Table

Structural features of the three major forms of DNA

Geometry Attribute	*A-DNA*	*B-DNA*	*Z-DNA*
Helix sense	right-handed	right-handed	left-handed
Repeating unit	1 bp	1 bp	2 bp
Rotation/bp	32.7°	35.9°	60°/2
bp/turn	11	10.5	12
Inclination of bp to axis	+19°	–1.2°	–9°
Rise/bp along axis	2.3 Å (0.23 nm)	3.32 Å (0.332 nm)	3.8 Å (0.38 nm)
Pitch/turn of helix	28.2 Å (2.82 nm)	33.2 Å (3.32 nm)	45.6 Å (4.56 nm)
Mean propeller twist	+18°	+16°	0°
Glycosyl angle	anti	anti	C: anti,G: syn
Sugar pucker	C3'-endo	C2'-endo	C: C2'-endo,G: C2'-exo
Diameter	23 Å (2.3 nm)	20 Å (2.0 nm)	18 Å (1.8 nm)
Sources:			

Grooves

Twin helical strands form the DNA backbone. Another double helix may be found by tracing the spaces, or grooves, between the strands. These voids are adjacent to the base pairs and may provide a binding site.

As the strands are not directly opposite each other, the grooves are unequally sized. One groove, the major groove, is 22 Å wide and the other, the minor groove, is 12 Å wide. The narrowness of the minor groove means that the edges of the bases are more accessible in the major groove. As a result, proteins like transcription factors that can bind to specific sequences in double-stranded DNA usually make contacts to the sides of the bases exposed in the major groove. This situation varies in unusual conformations of DNA within the cell, but the major and minor grooves are always named to reflect the differences in size that would be seen if the DNA is twisted back into the ordinary B form.

Non-double Helical Forms

Other non-double helical forms of DNA have been theorized, for example the side-by-side (SBS) geometry where the two strands do not wind around each other. Side-by-side models of DNA were proposed early in the history of molecular biology, but these were set aside in favor of the double-helical model due to X-ray crystallography of DNA duplexes and later the nucleosome core particle, as well as the discovery of topoisomerases. The state of current understanding in the field was aptly outlined in an exchange of correspondence in Current Science in 2004. Single-stranded nucleic acids do not adopt a helical formation, and are described by models such as the random coil or worm-like chain.

Bending

DNA is a relatively rigid polymer, typically modelled as a worm-like chain. It has three significant degrees of freedom; bending, twisting and compression, each of which cause particular limitations on what is possible with DNA within a cell. Twisting/torsional stiffness is important for the circularisation of DNA and the orientation of DNA bound proteins relative to each other and bending/axial stiffness is important for DNA wrapping and circularisation and protein interactions. Compression/extension is relatively unimportant in the absence of high tension.

Persistence Length/Axial Stiffness

Table: *Example sequences and their persistence lengths (B DNA)*

Sequence	*Persistence Length/base pairs*
Random	154±10
$(\mathbf{CA})_{repeat}$	133±10
$(\mathbf{CAG})_{repeat}$	124±10
$(\mathbf{TATA})_{repeat}$	137±10

DNA in solution does not take a rigid structure but is continually changing conformation due to thermal vibration and collisions with water molecules, which makes classical measures of rigidity impossible. Hence, the bending stiffness of DNA is measured by the persistence length, defined as:

"The length of DNA over which the time-averaged orientation of the polymer becomes uncorrelated by a factor of e".

This value may be directly measured using an atomic force microscope to directly image DNA molecules of various lengths. In an aqueous solution, the average persistence length is 46-50 nm or 140-150 base pairs (the diameter of DNA is 2 nm), although can vary significantly. This makes DNA a moderately stiff molecule.

The persistence length of a section of DNA is somewhat dependent on its sequence, and this can cause significant variation. The variation is largely due to base stacking energies and the residues which extend into the minor and major grooves.

Models for DNA Bending

Table: *Stacking stability of base steps (B DNA)*

Step	*Stacking ?G/kcal mol*
T A	-0.19
T G or C A	-0.55
C G	-0.91
A G or C T	-1.06
A A or T T	-1.11
A T	-1.34
G A or T C	-1.43
C C or G G	-1.44
A C or G T	-1.81
G C	-2.17

The entropic flexibility of DNA is remarkably consistent with standard polymer physics models such as the Kratky-Porod worm-like chain model. Consistent with the worm-like chain model is the observation that bending DNA is also described by Hooke's law at very small (sub-piconewton) forces. However for DNA segments less than the persistence length, the bending force is approximately constant and behaviour deviates from the worm-like chain predictions.

This effect results in unusual ease in circularising small DNA molecules and a higher probability of finding highly bent sections of DNA.

Bending Preference

DNA molecules often have a preferred direction to bend, ie. anisotropic bending. This is, again, due to the properties of the bases which make up the DNA sequence - a random sequence will have no preferred bend direction, i.e. isotropic bending.

Preferred DNA bend direction is determined by the stability of stacking each base on top of the next. If unstable base stacking steps are always found on one side of the DNA helix then the DNA will preferentially bend away from that direction. As bend angle increases then steric hindrances and ability to roll the residues relative to each other also play a role, especially in the minor groove. A and T residues will be preferentially be found in the minor grooves on the inside of bends. This effect is particularly seen in DNA-protein binding where tight DNA bending is induced, such as in nucleosome particles. See base step distortions above. DNA molecules with exceptional bending preference can become intrinsically bent. This was first observed in trypanosomatid kinetoplast DNA. Typical sequences which cause this contain stretches of 4-6 T and A residues separated by G and C rich sections which keep the A and T residues in phase with the minor groove on one side of the molecule. For example:

G A T T C C C A A A A A A T G T C A A A A A A A T A G G C A A A A A A T G C C A A A A A A A T C C C A A A C

The intrinsically bent structure is induced by the 'propeller twist' of base pairs relative to each other allowing unusual bifurcated Hydrogen-bonds between base steps. At higher temperatures this structure, and so thc intrinsic bend, is lost. All DNA which bends anisotropically has, on average, a longer persistence length and greater axial stiffness. This increased rigidity is required to prevent random bending which would make the molecule act isotropically.

Circularization

DNA circularization depends on both the axial (bending) stiffness and torsional (rotational) stiffness of the molecule. For a DNA molecule to successfully circularize it must be long enough to easily bend into the full circle and must have the correct number of bases so the ends are in the correct rotation to allow bonding to occur. The optimum length for circularization of DNA is around 400 base pairs (136 nm), with an integral number of turns of the DNA helix, i.e. multiples of 10.4 base pairs. Having a non integral number of turns presents a significant energy barrier for circularization, for example a 10.4 x 30 = 312 base pair molecule will circularize hundreds of times faster than 10.4 x 30.5 H" 317 base pair molecule.

Stretching

Longer stretches of DNA are entropically elastic under tension. When DNA is in solution, it undergoes continuous structural variations due to the energy available in the thermal bath of the solvent. This is due to the thermal vibration of the molecule combined with continual collisions with water molecules. For entropic reasons, more compact relaxed states are thermally accessible than stretched out states, and so DNA molecules are almost universally found in a tangled relaxed layouts. For this reason, a single molecule of DNA will stretch under a force, straightening it out.

Using optical tweezers, the entropic stretching behavior of DNA has been studied and analyzed from a polymer physics perspective, and it has been found that DNA behaves largely like the Kratky-Porod worm-like chain model under physiologically accessible energy scales. Under sufficient tension and positive torque, DNA is thought to undergo a phase transition with the bases splaying outwards and the phosphates moving to the middle. This proposed structure for overstretched DNA has been called "P-form DNA", in honor of Linus Pauling who originally presented it as a possible structure of DNA.

The mechanical properties DNA under compression have not been characterized due to experimental difficulties in preventing the polymer from bending under the compressive force.

Supercoiling and Topology

DNA supercoiling refers to the over- or under-winding of a DNA strand, and is an expression of the strain on the polymer. Supercoiling is important in a number of biological processes, such as compacting DNA. Additionally, certain enzymes such as topoisomerases are able to change DNA topology to facilitate functions such as DNA replication or transcription. Mathematical expressions are used to describe supercoiling by comparing different coiled states to relaxed B-form DNA. Supercoiled structure of circular DNA molecules with low writhe. Note that the helical nature of the DNA duplex is omitted for clarity.

Role of Supercoiling

In a "relaxed" double-helical segment of B-DNA, the two strands twist around the helical axis once every 10.4-10.5 base pairs of sequence. Adding or subtracting twists, as some enzymes can do, imposes strain. If a DNA segment under twist strain were closed into a circle by joining its two ends and then allowed to move freely, the circular DNA would contort into a new shape, such as a simple figure-eight. Such a contortion is a supercoil.

The simple figure eight is the simplest supercoil, and is the shape a circular DNA assumes to accommodate one too many or one too few helical twists. The two lobes of the figure eight will appear rotated either clockwise or counterclockwise with respect to one another, depending on whether the helix is over or underwound. For each additional helical twist being accommodated, the lobes will show one more rotation about their axis.

The noun form "supercoil" is rarely used in the context of DNA topology. Instead, global contortions of a circular DNA, such as the rotation of the figure-eight lobes above, are referred to as writhe. The above example illustrates that twist and writhe are interconvertible. "Supercoiling" is an abstract mathematical property representing the sum of twist and writhe.

The twist is the number of helical turns in the DNA and the writhe is the number of times the double helix crosses over on itself (these are the supercoils).

Extra helical twists are positive and lead to positive supercoiling, while subtractive twisting causes negative supercoiling. Many topoisomerase enzymes sense supercoiling and either generate or dissipate it as they change DNA topology. DNA of most organisms is negatively supercoiled.

In part because chromosomes may be very large, segments in the middle may act as if their ends are anchored. As a result, they may be unable to distribute excess twist to the rest of the chromosome or to absorb twist to recover from underwinding—the segments may become supercoiled, in other words. In response to supercoiling, they will assume an amount of writhe, just as if their ends were joined.

Supercoiled DNA forms two structures; a plectoneme or a toroid, or a combination of both. A negatively supercoiled DNA molecule will produce either a one-start left-handed helix, the toroid, or a two-start right-handed helix with terminal loops, the plectoneme. Plectonemes are typically more common in nature, and this is the shape most bacterial plasmids will take.

For larger molecules it is common for hybrid structures to form - a loop on a toroid can extend into a plectoneme. If all the loops on a toroid extend then it becomes a branch point in the plectonemic structure.

Occurrence of DNA Supercoiling

DNA supercoiling is important for DNA packaging within all cells. Because the length of DNA can be thousands of times that of a cell, packaging this genetic material into the cell or nucleus (in eukaryotes) is a difficult feat.

Supercoiling of DNA reduces the space and allows for much more DNA to be packaged. In prokaryotes, plectonemic supercoils are predominant, because of the circular chromosome and relatively small amount of genetic material. In eukaryotes, DNA supercoiling exists on many levels of both plectonemic and solenoidal supercoils, with the solenoidal supercoiling proving most effective in compacting the DNA. Solenoidal supercoiling is achieved with histones to form a 10 nm fiber. This fiber is further coiled into a 30 nm fiber, and further coiled upon itself numerous times more.

DNA packaging is greatly increased during nuclear division events such as mitosis or meiosis, where DNA must be compacted and segregated to daughter cells.

Condensins and cohesins are Structural Maintenance of Chromosome proteins that aid in the condensation of sister chromatids and the linkage of the centromere in sister chromatids. These SMC proteins induce positive supercoils. Supercoiling is also required for DNA/RNA synthesis. Because DNA must be unwound for DNA/RNA polymerase action, supercoils will result.

The region ahead of the polymerase complex will be unwound; this stress is compensated with positive supercoils ahead of the complex. Behind the complex, DNA is rewound and there will be compensatory negative supercoils. It is important to note that topoisomerases such as DNA gyrase (Type II Topoisomerase) play a role in relieving some of the stress during DNA/RNA synthesis.

Modeling Using Mathematics

DNA supercoiling can be described numerically by changes in the 'linking number' Lk. The linking number is the most descriptive property of supercoiled DNA. Lk_o, the number of turns in the relaxed (B type) DNA plasmid/molecule, is determined by dividing the total base pairs of the molecule by the relaxed bp/turn which, depending on reference is 10.4-10.5.

$$Lk_o = bp / 10.4$$

Lk is merely the number of crosses a single strand makes across the other in a planar projection. The topology of the DNA is described by the equation below in which the linking number is equivalent to the sum of TW, which is the number of twists or turns of the double helix, and Wr which is the number of coils or 'writhes'. If there is a closed DNA molecule, the sum of TW and Wr, or the linking number, does not change. However, there may be complementary changes in TW and Wr without changing their sum.

$$Lk = Tw + Wr$$

The change in the linking number, ÄLk, is the actual number of turns in the plasmid/molecule, Lk, minus the number of turns in the relaxed plasmid/molecule Lk_o.

$$\Delta Lk = Lk - Lk_o$$

If the DNA is negatively supercoiled $\Delta Lk < 0$. The negative supercoiling implies that the DNA is underwound.

A standard expression independent of the molecule size is the "specific linking difference" or "superhelical density" denoted σ. σ represents the number of turns added or removed relative to the total number of turns in the relaxed molecule/plasmid, indicating the level of supercoiling.

$$\sigma = \Delta Lk / Lk_o$$

The Gibbs free energy associated with the coiling is given by the equation below

$$\Delta G / N = 10RT\sigma$$

Examples

Since the linking number L of supercoiled DNA is the number of times the two strands are intertwined (and both strands remain covalently intact), L cannot change. The reference state (or parameter) L_0 of a circular DNA duplex is its relaxed state. In this state, its writhe W = 0. Since L = T + W, in a relaxed state T = L. Thus, if we have a 400 bp relaxed circular DNA duplex, L ~ 40 (assuming ~10 bp per turn in B-DNA). Then T ~ 40.

- Positively supercoiling:

  T = 0, W = 0, then L = 0

  T = +3, W = 0, then L = +3

  T = +2, W = +1, then L = +3

- Negatively supercoiling:

  T = 0, W = 0, then L = 0

  T = -3, W = 0, then L = -3

  T = -2, W = -1, then L = -3

Negative supercoils favor local unwinding of the DNA, allowing processes such as transcription, DNA replication, and recombination. Negative supercoiling is also thought to favour the transition between B-DNA and Z-DNA, and moderate the interactions of DNA binding proteins involved in gene regulation. The B form of the DNA helix twists 360° per 10.4-10.5 bp in the absence of torsional strain. But many molecular biological processes can induce torsional strain.

A DNA segment with excess or insufficient helical twisting is referred to, respectively, as positively or negatively "supercoiled". DNA in vivo is typically negatively supercoiled, which facilitates the unwinding (melting) of the double-helix required for RNA transcription. Within the cell most DNA is topologically restricted. DNA is typically found in closed loops (such as plasmids in prokaryotes) which are topologically closed, or as very long molecules whose diffusion coefficients produce effectively topologically closed domains. Linear sections of DNA are also commonly bound to proteins or physical structures (such as membranes) to form closed topological loops. Francis Crick was one of the first to propose the importance of linking numbers when considering DNA supercoils. In a paper published in 1976, Crick outlined the problem as follows:

In considering supercoils formed by closed double-stranded molecules of DNA certain mathematical concepts, such as the linking number and the twist, are needed. The meaning of these for a closed ribbon is explained and also that of the writhing number of a closed curve. Some simple examples are given, some of which may be relevant to the structure of chromatin.

Analysis of DNA topology uses three values:

L = linking number - the number of times one DNA strand wraps around the other. It is an integer for a closed loop and constant for a closed topological domain.

T = twist - total number of turns in the double stranded DNA helix. This will normally tend to approach the number of turns that a topologically open double stranded DNA helix makes free in solution: number of bases/10.5, assuming there are no intercalating agents (e.g., chloroquine) or other elements modifying the stiffness of the DNA.

W = writhe - number of turns of the double stranded DNA helix around the superhelical axis

$$L = T + W \text{ and } \Delta L = \Delta T + \Delta W$$

Any change of T in a closed topological domain must be balanced by a change in W, and vice versa. This results in higher order structure of DNA. A circular DNA molecule with a writhe of 0 will be circular. If the twist of this molecule is subsequently increased or decreased by supercoiling then the writhe will be appropriately altered, making the molecule undergo plectonemic or toroidal superhelical coiling. When the ends of a piece of double stranded helical DNA are joined so that it forms a circle the strands are topologically knotted. This means the single strands cannot be separated any process that does not involve breaking a strand (such as heating). The task of un-knotting topologically linked strands of DNA falls to enzymes known as topoisomerases. These enzymes are dedicated to un-knotting circular DNA by cleaving one or both strands so that another double or single stranded segment can pass through. This un-knotting is required for the replication of circular DNA and various types of recombination in linear DNA which have similar topological constraints.

The Linking Number Paradox

For many years, the origin of residual supercoiling in eukaryotic genomes remained unclear. This topological puzzle was referred to by some as the "linking number paradox". However, when experimentally determined structures of the nucleosome displayed an over-twisted left-handed wrap of DNA around the histone octamer, this "paradox" was considered to be solved by the scientific community.

Full genome sequencing

Full genome sequencing (FGS), also known as whole genome sequencing (WGS), complete genome sequencing, or entire genome sequencing, is a laboratory process that determines the complete DNA sequence of an organism's genome at a single time. This entails sequencing all of an organism's chromosomal DNA as well as DNA contained in the mitochondria and, for plants, in the chloroplast. Almost any biological sample—even a very small amount of DNA or ancient DNA—can provide the genetic material necessary for full genome sequencing. Such samples may include saliva, epithelial cells, bone marrow, hair (as long as the hair contains a hair follicle), seeds, plant leaves, or anything else that has DNA-containing cells. Because the sequence data that is produced can be quite large (for example, there are approximately six billion base pairs in each human diploid genome), genomic data is stored electronically and requires a large amount of computing power and storage capacity. Full genome sequencing would have been nearly impossible before the advent of the microprocessor, computers, and the Information Age. Unlike full genome sequencing, DNA profiling only determines the likelihood that genetic material came from a particular individual or group; it does not contain additional information on genetic relationships, origin or susceptibility to specific diseases. Also unlike full genome sequencing, SNP genotyping covers less than 0.1% of the genome. Almost all truly complete genomes are of microbes; the term "full genome" is thus sometimes used loosely to mean "greater than 95%". The remainder of this article focuses on nearly complete human genomes. In general, knowing the complete DNA sequence of an individual's genome does not, on its own, provide useful clinical information, but this may change over time as a large number of scientific studies continue to be published detailing clear associations between specific genetic variants and disease.

The first nearly complete human genomes sequenced were J. Craig Venter's (Caucasian at 7.5-fold average coverage), James Watson's (Caucasian male at 7.4-fold), a Han Chinese (YH at 36-fold), a Yoruban from Nigeria (at 30-fold), a female leukemia patient (at 33 and 14-fold coverage for tumor and normal tissues), and Seong-Jin Kim (Korean at 29-fold). There are currently over 60 nearly complete human genomes publicly available. Steve Jobs also had his genome sequenced for $100,000. Commercialization of full genome sequencing is in an early stage and growing rapidly.

Early Techniques

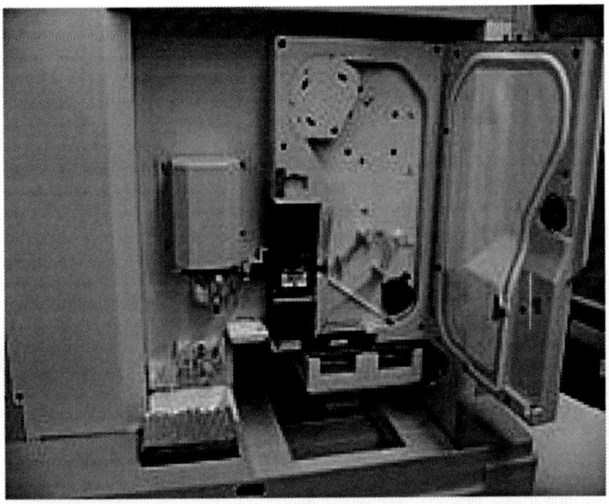

Figure: *An ABI PRISM 3100 Genetic Analyzer. Such capillary sequencers automated the early efforts of sequencing genomes.*

Sequencing of nearly an entire human genome was first accomplished in 2000 partly through the use of shotgun sequencing technology. While full genome shotgun sequencing for small (4000–7000 base pair) genomes was already in use in 1979, broader application benefited from pairwise end sequencing, known colloquially as double-barrel shotgun sequencing. As sequencing projects began to take on longer and more complicated genomes, multiple groups began to realize that useful information could be obtained by sequencing both ends of a fragment of DNA. Although sequencing both ends of the same fragment and keeping track of the paired data was more cumbersome than sequencing a single end of two distinct fragments, the knowledge that the two sequences were oriented in opposite directions and were about the length of a fragment apart from each other was valuable in reconstructing the sequence of the original target fragment.

The first published description of the use of paired ends was in 1990 as part of the sequencing of the human HPRT locus, although the use of paired ends was limited to closing gaps after the application of a traditional shotgun sequencing approach. The first theoretical description of a pure pairwise end sequencing strategy, assuming fragments of constant length, was in 1991. In 1995 Roach et al. introduced the innovation of using fragments of varying sizes, and demonstrated that a pure pairwise end-sequencing strategy would be possible on large targets. The strategy was subsequently adopted by The Institute for Genomic Research (TIGR) to sequence the entire genome of the bacterium Haemophilus influenzae in 1995, and then by Celera Genomics to sequence the entire fruit fly genome in 2000, and subsequently the entire human genome. Applied Biosystems, now called Life Technologies, manufactured the automated capillary sequencers utilized by both Celera Genomics and The Human Genome Project. While capillary sequencing was the first approach to successfully sequence a nearly full human genome, it is still too expensive and takes too long for commercial purposes. Because of this, shotgun sequencing technology, even though it is still relatively 'new', since 2005 is being displaced by technologies like pyrosequencing, SMRT sequencing, and nanopore technology.

Current Research

One possible way to accomplish the cost-effective high-throughput sequencing necessary to accomplish full genome sequencing is by using nanopore technology, which is a patented technology held by Harvard University and Oxford Nanopore Technologies and licensed to biotechnology companies. To facilitate their full genome sequencing initiatives, Illumina licensed nanopore sequencing technology from Oxford Nanopore Technologies and Sequenom licensed the technology from Harvard University. Another possible way to accomplish cost-effective high-throughput sequencing is by utilizing fluorophore technology. Pacific Biosciences is currently using this approach in their SMRT (single molecule real time) DNA sequencing technology. Complete Genomics has developed DNA Nanoball (DNB) technology that arranges DNA on self-assembling arrays. Complete Genomics' sequencing technology combines its DNB arrays with its proprietary cPAL™ read technology. Pyrosequencing is a method of DNA sequencing based on the sequencing by synthesis principle. The technique was developed by Pål Nyrén and his student Mostafa Ronaghi at the Royal Institute of Technology in Stockholm in 1996, and is currently being used by 454 Life Sciences in their effort to deliver an affordable, fast and highly accurate full genome sequencing platform.

Commercialization

A number of public and private companies are competing to develop a full genome sequencing platform that is commercially robust for both research and clinical use, including Illumina, Knome, Sequenom, 454 Life Sciences, Pacific Biosciences, Complete Genomics, Intelligent Bio-Systems, Genome Corp., ION Torrent Systems, Helicos Biosciences, and BioNanomatrix. These companies are heavily financed and backed by venture capitalists, hedge funds, and investment banks.

Incentive

In October 2006, the X Prize Foundation, working in collaboration with the J. Craig Venter Science Foundation, established the Archon X Prize for Genomics, intending to award US$10 million to "the first Team that can build a device and use it to sequence 100 human genomes within 10 days or less, with an accuracy of no more than one error in every 100,000 bases sequenced, with sequences accurately covering at least 98% of the genome, and at a recurring cost of no more than US$10,000 per genome." An error rate of 1 in 100,000 bases, out of a total of six billion bases in the human diploid genome, would mean about 60,000 errors per genome. The error rates required for widespread clinical use, such as Predictive Medicine is currently set by over 1,400 clinical single gene sequencing tests (for example, errors in BRCA1 gene for breast cancer risk analysis). As of July 2011, the Archon X Prize for Genomics remains unclaimed.

2007

In 2007, Applied Biosystems started selling a new type of sequencer called SOLiD System. The technology allowed users to sequence 60 gigabases per run.

2009

In early February 2009, Complete Genomics released a full sequence of a human genome that was sequenced using their service. The data indicates that Complete Genomics' full genome sequencing service accuracy is just under 99.999%, meaning that just one in every one hundred thousand variants was called incorrectly. This means that their full sequence of the human genome will contain approximately 80,000–100,000 false positive errors in each genome. However, this accuracy rate was based on Complete Genomics' sequence that was completed utilizing a 90× depth of coverage (each base in the genome was sequenced 90 times) while their commercialized sequence is reported to be only 40×, so the accuracy may be substantially lower unless they can find some way to improve it before their first service release planned for the summer 2009. This accuracy rate may be acceptable for research purposes, and clinical use would require confirmation by other methods of any reportable alleles. Complete Genomics announced in Dec. 2010 that for the last 500 complete human genomes that it had sequenced, an average of over 98 percent of the genome was read at 10-fold or greater coverage. In addition, its software made high confidence calls of an average of over 95 percent of the genome and over 94 percent of the exome.

In March 2009, it was announced that Complete Genomics has signed a deal with the Broad Institute to sequence cancer patients' genomes and will be sequencing five full genomes to start. In April 2009, Complete Genomics announced that it plans to sequence 1,000 full genomes between June 2009 and the end of the year and that they plan to be able to sequence one million full genomes per year by 2013.

Complete Genomics sequenced 50 genomes in 2009. Since then, it has significantly increased the throughout in its genome sequencing center and was able to sequence and analyze 300 complete human genomes in Q3 2010. Complete Genomics plans to officially launch in June 2009, although it is unknown if their lab will have received CLIA-certification by that time. Complete Genomics announced its R&D human genome sequencing service in October 2008 and its commercial sequencing service in May 2010. The company does not produce clinical data and as such its genome center does not require CLIA certification.

In June 2009, NABsys announced their goal of full genome sequencing for under US$100 per genome with a turnaround time of less than 15 minutes.

In June 2009, Illumina announced that they were launching their own Personal Full Genome Sequencing Service at a depth of 30× for US$48,000 per genome. This is still expensive for widespread consumer use, but the price may decrease substantially over the next few years as they realize economies of scale and given the competition with other companies such as Complete Genomics.

Jay Flatley, Illumina's President and CEO, stated that "during the next five years, perhaps markedly sooner," the price point for full genome sequencing will fall from US$48,000 to under US$1,000. Illumina has already signed agreements to supply full genome sequencing services to multiple direct-to-consumer personal genomics companies.

In August 2009, the founder of Helicos Biosciences, Dr. Stephen Quake, stated that using the company's Heliscope Single Molecule Sequencer he sequenced his own full genome for less than US$50,000. He stated that he expects the cost to decrease to the US$1,000 range within the next two to three years.

In August 2009, Pacific Biosciences secured an additional US$68 million in new financing, bringing their total capitalization to US$188 million. Pacific Biosciences said they are going to use this additional investment in order to prepare for the upcoming launch of their full genome sequencing service in 2010. Complete Genomics followed by securing another US$45 million in a fourth round venture funding during the same month. Complete Genomics has also made the claim that it will sequence 10,000 full genomes by the end of 2010. Since then, it has significantly increased the throughout in its genome sequencing center and was able to sequence and analyze 300 complete human genomes in Q3 2010.

GE Global Research is also part of this race to commercialize full genome sequencing as they have been working on creating a service that will deliver a full genome for US$1,000 or less. In September 2009, the President of Halcyon Molecular announced that they will be able to provide full genome sequencing in under 10 minutes for less than US$100 per genome. This is, to date, the most ambitious promise of any full genome sequencing company.

In October 2009, IBM announced that they were also in the heated race to provide full genome sequencing for under US$1,000, with their ultimate goal being able to provide their service for US$100 per genome. IBM's full genome sequencing technology, which uses nanopores, is known as the "DNA Transistor".

In November 2009, Complete Genomics published a peer-reviewed paper in Science demonstrating its ability to sequence a complete human genome for US$1,700. If true, this would mean the cost of full genome sequencing has come down exponentially within just a single year from around US$100,000 to US$50,000 and now to US$1,700. This consumables cost was clearly detailed in the Science paper. However, Complete Genomics has previously released statements that it was unable to follow through on. For example, the company stated it would officially launch and release its service during the "summer of 2009", provide a "US$5,000" full genome sequencing service by the "summer of 2009", and "sequence 1,000 genomes between June 2009 and the end of 2009" – all of which, as of November 2009, have not yet occurred. Complete Genomics launched its R&D human genome sequencing service in October 2008 and its commercial service in May 2010. The company sequenced 50 genomes in 2009. Since then, it has significantly increased the throughput of its genome sequencing factory and was able to sequence and analyze 300 genomes in Q3 2010.

Also in November 2009, Complete Genomics announced that it was beginning a large-scale human genome sequencing study of Huntington's disease (up to 100 genomes) with the Institute for Systems Biology.

2010

In March 2010 Researchers from the Medical College of Wisconsin announced the first successful use of Genome Wide sequencing to change the treatment of a patient. This story was later retold in a pulitzer prize winning article and touted as a significant accomplishment in Nature and by the director of the NIH in presentations at congress.

In March 2010, Pacific Biosciences said they have raised more than US$256 million in venture capital money and that they will be shipping their first ten full genome sequencing machines by the end of 2010. The company reported that the market initially will be researchers and academic institutions and then will rapidly turn into clinical applications that will be applicable to every single person in the world. Pacific Biosciences also stated that their second-generation machine, which is scheduled for release in 2015, will be capable of providing a full genome sequence for a person in just 15 minutes for less than US$100. Several other technologies have similar goals. Meanwhile, full genome sequencing might revolutionize medicine at even current prices by providing a clinician with a full genome for each one of his or her patients. However, some critics have stated that even if they are supplied with a full genome sequence of a patient, they would not know how to analyze or make use of that data. Since then, new resources have begun to address this.

Also in March 2010, Complete Genomics' customers began publishing papers describing research breakthroughs that they have made using data it has provided. Examples included the Institute for Systems Biology's project to sequence a family of four and verify the gene responsible for Miller Syndrome, a rare craniofacial disorder[x] and Genentech's work to sequence and compare a patient's primary lung tumor and adjacent normal tissue[y]. In June 2010, Illumina lowered the cost of its individual sequencing service to US$19,500 from US$48,000. The company is offering a discounted price of US$9,500 for people with serious medical conditions who could potentially benefit from having their genomes decoded.

2011

Knome provides full genome (98% genome) sequencing services for US$39,500 for whole genome sequencing and interpretation for consumers. It's US$29,500 for whole genome sequencing and analysis for researchers depending on their requirements. Complete Genomics charges approximately US$10,000 to sequence a complete human genome and offers discounts for large orders. This service includes sample quality control, library preparation, sequencing, mapping, assembly and data analysis.

In May 2011, Illumina announced that it had lowered its Full Genome Sequencing service to US$5,000 per human genome and US$4,000 per genome for bulk orders of 50 or more.

Helicos Biosciences, Pacific Biosciences, Complete Genomics, Illumina, Sequenom, ION Torrent Systems, Halcyon Molecular, NABsys, IBM, and GE Global appear to all be going head to head in the race to commercialize full genome sequencing.

Disruptive Technology

Full genome sequencing provides information on a genome that is orders of magnitude larger than that provided by the current leader in genotyping technology, DNA arrays. For humans, DNA arrays currently provide genotypic information on up to one million genetic variants, while full genome sequencing will provide information on all six billion bases in the human genome, or 3,000 times more data. Because of this, full genome sequencing is considered disruptive to the DNA array markets as the accuracy of both range from 99.98% to 99.999% (in non-repetitive DNA regions) and their consumables cost of US$5000 per 6 billion base pairs is competitive (for some applications) with DNA arrays (US$500 per 1 million basepairs).

Agilent, another established DNA array manufacturer, is working on targeted (selective region) genome sequencing technologies. It is thought that Affymetrix, the pioneer of array technology in the 1990s, has fallen behind due to significant corporate and stock turbulence and is currently not working on any known full genome sequencing approach.

It is unknown what will happen to the DNA array market once full genome sequencing becomes commercially widespread, especially as companies and laboratories providing this disruptive technology start to realize economies of scale. It is postulated, however, that this new technology may significantly diminish the total market size for arrays and any other sequencing technology once it becomes commonplace for individuals and newborns to have their full genomes sequenced.

Sequencing Versus Analysis

Full genome sequencing provides raw data on all six billion letters in an individual's DNA. However, it does not provide an analysis of what that data means or how that data can be utilized in various clinical applications, such as in medicine to help prevent disease. As of 2010 the companies that are working on providing full genome sequencing provide clinical CLIA certified data (Illumina) and analytical services for the interpretation of the full genome data (Knome) With only one institition offering sequencing and analysis in a clinical setting. Nevertheless there is plenty of room for researchers or companies to improve such analyses and make it useful to physicians and patients.

Societal Impact

Inexpensive, time-efficient full genome sequencing will be a major accomplishment not only for the field of Genomics, but for the entire human civilization because, for the first time, individuals will be able to have their entire genome sequenced. Utilizing this information, it is speculated that health care professionals, such as physicians and genetic counselors, will eventually be able to use genomic information to predict what diseases a person may get in the future and attempt to either minimize the impact of that disease or avoid it altogether through the implementation of personalized, preventive medicine.

Full genome sequencing will allow health care professionals to analyze the entire human genome of an individual and therefore detect all disease-related genetic variants, regardless of the genetic variant's prevalence or frequency. This will enable the rapidly emerging medical fields of Predictive Medicine and Personalized Medicine and will mark a significant leap forward for the clinical genetic revolution. Full genome sequencing is clearly of great importance for research into the basis of genetic disease and has shown significant benefit to a subset of individuals with rare disease in the clinical setting.

Illumina's CEO, Jay Flatley, stated in February 2009 that "A complete DNA read-out for every newborn will be technically feasible and affordable in less than five years, promising a revolution in healthcare" and that "by 2019 it will have become routine to map infants' genes when they are born." This potential use of genome sequencing is highly controversial, as it runs counter to established ethical norms for predictive genetic testing of asymptomatic minors that have been well established in the fields of medical genetics and genetic counseling.

The traditional guidelines for genetic testing have been developed over the course of several decades since it first became possible to test for genetic markers associated with disease, prior to the advent of cost-effective, comprehensive genetic screening. It is established that norms, such as in the sciences and the field of genetics, are subject to change and evolve over time. It is unknown whether traditional norms practiced in medical genetics today will be altered by new technological advancements such as full genome sequencing.

Today, parents have the legal authority to obtain testing of any kind for their children. Currently available newborn screening for childhood diseases allows detection of rare disorders that can be prevented or better treated by early detection and intervention. Specific genetic tests are also available to determine an etiology when a child's symptoms appear to have a genetic basis. Full genome sequencing, in addition has the potential to reveal a large amount of information (such as carrier status for autosomal recessive disorders, genetic risk factors for complex adult-onset diseases, and other predictive medical and non-medical information) that is currently not completely understood, may not be clinically useful to the child during childhood, and may not necessarily be wanted by the individual upon reaching adulthood.

Despite the benefits of predicting disease risk in childhood, genetic testing also introduces potential harms (such as discovery of non-paternity, genetic discrimination, and psychological impacts). The established ethical guidelines for predictive genetic testing of asymptomatic minors thus has more to do with protecting this vulnerable population and preserving the individual's privacy and autonomy to know or not to know their genetic information, than with the technology that makes this possible. While parents may have legal authority to obtain such testing, the mainstream opinion of professional medical genetics societies is that presymptomatic testing should be offered to minors only when they are competent to understand the relevancy of genetic screening so as to allow them to participate in the decision about whether or not it is appropriate for them.

Gene-centric View of Evolution

The gene-centered view of evolution, gene selection theory or selfish gene theory holds that evolution occurs through the differential survival of competing genes, increasing the frequency of those alleles whose phenotypic effects successfully promote their own propagation, with gene defined as "not just one single physical bit of DNA [but] all replicas of a particular bit of DNA distributed throughout the world". The proponents of this viewpoint argue that, since heritable information is passed from generation to generation almost exclusively by genetic material, natural selection and evolution are best considered from the perspective of genes.

This is in contrast to the organism-centered viewpoint adopted historically by biologists. Proponents of the gene-centered viewpoint argue that it permits understanding of diverse phenomena such as altruism and intragenomic conflict that are otherwise difficult to explain from an organism-focused perspective.

The gene-centered view of evolution is a synthesis of the theory of evolution by natural selection, the particulate inheritance theory and the non-transmission of acquired characters. It states that those genes whose phenotypic effects successfully promote their own propagation will be favorably selected in detriment to their competitors. This process produces adaptations for the benefit of genes that promote the reproductive success of the organism, or of other organisms containing the same gene (kin altruism and green-beard effects), or even only its own propagation in detriment to the other genes of the genome (intragenomic conflict).

Introduction

The gene-centered view of evolution is a different way of looking at the basis of evolutionary development. It turns the whole solution of evolution inside-out for the purpose of examination. What this new perspective reveals is a more easily understood model for the evolution of social characteristics such as selfishness and altruism that much of the study of evolution, caught up in the survival of the individual organisms, overlooks.

Acquired Characteristics are not Inherited

Discoveries in science such as the formulation of the central dogma of molecular biology made it clear that the inheritance of acquired characters was not an evolutionary factor in a physical sense and identified genes as lasting entities that survive through many generations. Maynard Smith summarized the issue:

"If the central dogma is true, and if it is also true that nucleic acids are the only means whereby information is transmitted between generations, this has crucial implications for evolution. It would imply that all evolutionary novelty requires changes in nucleic acids, and that these changes – mutations – are essentially accidental and non-adaptive in nature. Changes elsewhere – in the egg cytoplasm, in materials transmitted through the placenta, in the mother's milk – might alter the development of the child, but, unless the changes were in nucleic acids, they would have no long-term evolutionary effects."

The rejection of the inheritance of acquired characters, combined with the classical mathematical evolutionary biology developed by Ronald Fisher, J. B. S. Haldane and Sewall Wright, paved the way to the formulation of the selfish-gene theory. For cases where environment can influence heredity.

The Gene as the Unit of Selection

The view of the gene as the unit of selection was developed mainly in the works W. D. Hamilton, Colin Pittendrigh and George C. Williams. It was later popularized by Richard Dawkins in his books The Selfish Gene (1976) and The Extended Phenotype (1982).

According to Williams' 1966 book Adaptation and Natural Selection,

"[t]he essence of the genetical theory of natural selection is a statistical bias in the relative rates of survival of alternatives (genes, individuals, etc.). The effectiveness of such bias in producing adaptation is contingent on the maintenance of certain quantitative relationships among the operative factors. One necessary condition is that the selected entity must have a high degree of permanence and a low rate of endogenous change, relative to the degree of bias (differences in selection coefficients)."

Williams argued that "[t]he natural selection of phenotypes cannot in itself produce cumulative change, because phenotypes are extremely temporary manifestations". Each phenotype is the unique product of the interaction between genome and environment. It does not matter how fit and fertile a phenotype is, it will eventually be destroyed and will never be duplicated.

Since 1954, it has been known that DNA is the main physical substrate to genetic information, and it is capable of high-fidelity replication through many generations. So, a particular sequence of DNA can have a high permanence and a low rate of endogenous change. In normal sexual reproduction, an entire genome is the unique combination of father's and mother's chromosomes produced at the moment of fertilization. It is generally destroyed with its organism, because "meiosis and recombination destroy genotypes as surely as death". Only half of it is transmitted to each descendant due to independent segregation.

The gene as an informational entity persists for an evolutionarily significant span of time through a lineage of many physical copies.

In his book River out of Eden, Dawkins coins the phrase God's utility function to explain his view on genes as units of selection. He uses this phrase as a synonym of the "meaning of life" or the "purpose of life". By rephrasing the word purpose in terms of what economists call a utility function, meaning "that which is maximized", Dawkins attempts to reverse-engineer the purpose in the mind of the Divine Engineer of Nature, or the Utility Function of God. Finally, Dawkins argues that it is a mistake to assume that an ecosystem or a species as a whole exists for a purpose. He writes that it is incorrect to suppose that individual organisms lead a meaningful life either; in nature, only genes have a utility function – to perpetuate their own existence with indifference to great sufferings inflicted upon the organisms they build, exploit and discard.

Organisms as Vehicles

Genes are not naked in the world. They are usually packed together inside a genome, which is itself contained inside an organism. Genes group together into genomes because "genetic replication makes use of energy and substrates that are supplied by the metabolic economy in much greater quantities than would be possible without a genetic division of labour". They build vehicles to promote their mutual interests of jumping into the next generation of vehicles. As Dawkins puts it, organisms are the "survival machines" of genes.

The phenotypic effect of a particular gene is contingent on its environment, including the fellow genes constituting with it the total genome. A gene never has a fixed effect, so how is it possible to speak of a gene for long legs? It is because of the phenotypic differences between alleles. One may say that one allele, all other things being equal or varying within certain limits, causes greater legs than its alternative. This difference enables the scrutiny of natural selection. “A gene can have multiple phenotypic effects, each of which may be of positive, negative or neutral value. It is the net selective value of a gene’s phenotypic effect that determines the fate of the gene”. For instance, a gene can cause its bearer to have greater reproductive success at a young age, but also cause a greater likelihood of death at a later age. If the benefit outweighs the harm, averaged out over the individuals and environments in which the gene happens to occur, then phenotypes containing the gene will generally be positively selected and thus the abundance of that gene in the population will increase. Even so, it becomes necessary to model the genes in combination with their vehicle as well as in combination with the vehicle’s environment.

Selfish-gene Theory

The selfish-gene theory of natural selection can be restated as follows:

"Genes do not present themselves naked to the scrutiny of natural selection, instead they present their phenotypic effects. [...] Differences in genes give rise to differences in these phenotypic effects. Natural selection acts on the phenotypic differences and thereby on genes. Thus genes come to be represented in successive generations in proportion to the selective value of their phenotypic effects." The result is that “the prevalent genes in a sexual population must be those that, as a mean condition, through a large number of genotypes in a large number of situations, have had the most favourable phenotypic effects for their own replication”. In other words, we expect selfish genes (“selfish” meaning that it promotes its own survival without necessarily promoting the survival of the organism, group or even species). This theory implies that adaptations are the phenotypic effects of genes to maximize their representation in future generations. An adaptation is maintained by selection if it promotes genetic survival directly, or else some subordinate goal that ultimately contributes to successful reproduction.

Individual Altruism, Genetic Egoism

The gene is a unit of hereditary information that exists in many physical copies in the world, and which particular physical copy will be replicated and originate new copies does not matter from the gene’s point of view. A selfish gene could be favored by selection by producing altruism among organisms containing it. The idea is summarized as follows:

If a gene copy confers a benefit B on another vehicle at cost C to its own vehicle, its costly action is strategically beneficial if $pB > C$, where p is the probability that a copy of the gene is present in the vehicle that benefits. Actions with substantial costs therefore require significant values of p. Two kinds of factors ensure high values of p: relatedness (kinship) and recognition (green beards).

A gene in a somatic cell of an individual may forego replication to promote the transmission of its copies in the germ line cells. It ensures the high value of p = 1 due to their constant contact and their common origin from the zygote.

The kin selection theory predicts that a gene may promote the recognition of kinship by historical continuity: a mammalian mother learns to identify her own offspring in the act of giving birth; a male preferentially directs resources to the offspring of mothers with whom he has copulated; the other chicks in a nest are siblings; and so on.

The expected altruism between kin is calibrated by the value of p, also known as the coefficient of relatedness. For instance, an individual has a p = 1/2 in relation to his brother, and p = 1/8 to his cousin, so we would expect, ceteris paribus, greater altruism among brothers than among cousins. In this vein, geneticist J.B.S. Haldane famously joked, "Would I lay down my life to save my brother? No, but I would to save two brothers or eight cousins". However, examining the human propensity for altruism, kin selection theory seems incapable of explaining cross-familiar, cross-racial and even cross-species acts of kindness.

Green-beard Effect

Green-beard effects gained their name from a thought-experiment of Richard Dawkins, who considered the possibility of a gene that caused its possessors to develop a green beard and to be nice to other green-bearded individuals. Since then, 'green-beard effect' has come to refer to forms of genetic self-recognition in which a gene in one individual might direct benefits to other individuals that possess the gene. Such genes are essentially especially selfish, benefiting themselves regardless of the fates of their vehicles.

All kinds of Altruism

Kindness

On the other hand, a single trait, group reciprocal kindness, is capable of explaining the vast majority of altruism that is generally accepted as "good" by modern societies. Imagine a green-bearding behavioral trait whose recognition does not depend on the recognition of some external feature such as beard color, but relies on recognition of the behavior itself. Imagine now that the behavior is altruistic. The success of such a trait in sufficiently intelligent and undeceived organisms is implicit. Moreover, the existence of such a trait predicts a tendency for kindness to unrelated organisms that are apparently kind, even if the organisms are of a completely different species. Moreover, the gene need not be exactly the same, so long as the effect is similar. Multiple versions of the gene—or even meme—would have virtually the same effect in a sort of symbiotic green-bearding cycle of altruism.

Deceit

Whenever recognition plays a role in evolution, so does deception. Just like the harmless lizard that has evolved a pattern that mimics its poisonous cousin and therefore tricks predators, the selfish creature may pretend to be kind by "growing a green beard" (whatever that green beard may be). Thus green-bearding and the selfish-gene theory also give rise to an explanation for the evolution of lies and deceit, characteristics that do not benefit the population as a whole.

Intragenomic Conflict

As genes are capable of producing individual altruism, they are capable of producing conflict among genes inside the genome of one individual. This phenomenon was called intragenomic conflict and arises when one gene promotes its own replication in detriment to other genes in the genome. The classic example is segregation distorter genes that cheat during meiosis or gametogenesis and end up in more than half of the functional gametes. These genes persist even resulting in reduced fertility. Egbert Leigh compared the genome to "a parliament of genes: each acts in its own self-interest, but if its acts hurt the others, they will combine together to suppress it" to explain the relative low occurrence of intragenomic conflict.

Criticism

Prominent opponents of this gene-centric view of evolution include evolutionary biologist Ernst Mayr, paleontologist Stephen Jay Gould, biologist and anthropologist David Sloan Wilson and philosopher Elliott Sober.

Writing in the New York Review of Books, Gould has characterized the gene-centered perspective as confusing book-keeping with causality. Gould views selection as working on many levels, and has called attention to a hierarchical perspective of selection. Gould also called the claims of Selfish Gene "strict adaptationism", "ultra-Darwinism", and "Darwinian fundamentalism", describing them as excessively "reductionist". He saw the theory as leading to a simplistic "algorithmic" theory of evolution, or even to the re-introduction of a teleological principle. Mayr went so far as to say "Dawkins' basic theory of the gene being the object of evolution is totally non-Darwinian".

Gould also addressed the issue of selfish genes in his essay 'Caring groups and selfish genes'. Gould acknowledged that Dawkins was not imputing conscious action to genes, but simply using a shorthand metaphor commonly found in evolutionary writings. To Gould, the fatal flaw was that "no matter how much power Dawkins wishes to assign to genes, there is one thing that he cannot give them – direct visibility to natural selection". Rather, the unit of selection is the phenotype, not the genotype, because it is phenotypes which interact with the environment at the natural-selection interface. So, in Kim

Sterelny's summation of Gould's view, "gene differences do not cause evolutionary changes in populations, they register those changes". Richard Dawkins replied to this criticism in a later book, The Extended Phenotype, that Gould confused particulate genetics with particulate embryology, stating that genes do "blend", as far as their effects on developing phenotypes are concerned, but that they do not blend as they replicate and recombine down the generations.

Since Gould's death in 2002, Niles Eldredge has continued with counter-arguments to gene-centered natural selection. Eldredge notes that in Dawkins' book A Devil's Chaplain, which was published just before Eldredge's book, "Richard Dawkins comments on what he sees as the main difference between his position and that of the late Stephen Jay Gould. He concludes that it is his own vision that genes play a causal role in evolution", while Gould (and Eldredge) "sees genes as passive recorders of what worked better than what".

Price Equation

The Price equation (also known as Price's equation) is a covariance equation which is a mathematical description of evolution and natural selection. The Price equation was derived by George R. Price, working in London to re-derive W.D. Hamilton's work on kin selection. The Price equation also has applications in economics.

Price's equation is a theorem; it is a statement of mathematical fact between certain variables, and its value lies in the insight gained by assigning certain values encountered in evolutionary genetics to the variables. For example, it may be known that every pair of birds has two offspring. This can be encoded into the Price equation by saying that the fitness quantity of each individual has a value of 1, where "fitness" is defined to be the number of offspring from each individual. If it is also assumed that each offspring inherits the fitness of its parent, then:

- Every individual in every generation of the population will have equal fitness.
- The population size will double from one generation to the next generation.
- Genes from an individual in one population will be equally represented in the next population. That is, the proportion of the population with those genes (e.g., 1/n for a population of n individuals) will be the same in subsequent generations (e.g., 2/(2n) for the population of 2n offspring).

These three points are mathematical facts following from knowledge of the fitness of every individual in one generation and how that fitness changes from parent to offspring. The Price equation provides the third bullet in more complicated situations where fitness may change from parent to offspring and certain individuals in a population may have a differential fitness advantage (i.e., individuals with certain traits will have more offspring than individuals with other traits). That is, the Price equation describes how the proportion of traits expressed in one generation changes over generational time. As the Price equation is a model of natural selection, it generally predicts that traits that yield a higher number of offspring than other traits will be favored in future generations.

Main Figures in Selection Debate

Besides Richard Dawkins and George C. Williams, other biologists and philosophers have expanded and refined the selfish-gene theory, such as John Maynard Smith, George R. Price, Robert Trivers, David Haig, Helena Cronin, David Hull, Philip Kitcher and Daniel C. Dennett. Individuals opposing this gene-centric view include Ernst Mayr, Stephen Jay Gould, David Sloan Wilson and philosopher Elliott Sober. Proponents of multilevel selection (MLS) include E.O. Wilson, David Sloan Wilson, Elliott Sober, Richard E. Michod and Samir Okasha.

Gene Dosage

Gene dosage is the number of copies of a gene present in a cell or nucleus. An increase in gene dosage can cause higher levels of gene product if the gene is not subject to regulation from elsewhere in the body.

Copy-number Variation

Copy-number variations (CNVs)—a form of structural variation—are alterations of the DNA of a genome that results in the cell having an abnormal number of copies of one or more sections of the DNA. CNVs correspond to relatively large regions of the genome that have been deleted (fewer than the normal number) or duplicated (more than the normal number) on certain chromosomes. For example, the chromosome that normally has sections in order as A-B-C-D might instead have sections A-B-C-C-D (a duplication of "C") or A-B-D (a deletion of "C").

This variation accounts for roughly 12% of human genomic DNA and each variation may range from about one kilobase (1,000 nucleotide bases) to several megabases in size. CNVs contrast with single-nucleotide polymorphisms (SNPs), which affect only one single nucleotide base.

Sources

CNVs may either be inherited or caused by de novo mutation. A recently proposed mechanism for the cause of some CNVs was fork stalling and template switching, a replication misstep. However, this model was subsequently superseded by microhomology-mediated break-induced replication (MMBIR).

CNVs can be caused by structural rearrangements of the genome such as deletions, duplications, inversions, and translocations. Low copy repeats (LCRs), which are region-specific repeat sequences, are susceptible to such genomic rearrangements resulting in CNVs. Factors such as size, orientation, percentage similarity and the distance between the copies influence the susceptibility of LCRs to genomic rearrangement. Segmental Duplications (SDs) map near ancestral duplication sites in a phenomenon called duplication shadowing which describes the observation of a ~10 fold increased probability of duplication in regions flanking duplications versus other random regions.

Identification

Copy number variation can be discovered by cytogenetic techniques such as fluorescent in situ hybridization, comparative genomic hybridization, array comparative genomic hybridization, and by virtual karyotyping with SNP arrays. Recent advances in DNA sequencing technology have further enabled the identification of CNVs by next-generation sequencing.

CNVs can be limited to a single gene or include a contiguous set of genes. CNVs can result in having either too many or too few of the dosage-sensitive genes, which may be responsible for a substantial amount of human phenotypic variability, complex behavioral traits and disease susceptibility.

In certain cases, such as rapidly growing Escherichia coli cells, the gene copy number can be 4-fold greater for genes located near the origin of DNA replication, rather than at the terminus of DNA replication. Elevating the gene copy number of a particular gene can increase the expression of the protein that it encodes.

Prevalence in Humans

The fact that DNA copy number variation is a widespread and common phenomenon among humans was first uncovered following the completion of the human genome project. It is estimated that approximately 0.4% of the genomes of unrelated people typically differ with respect to copy number. De novo CNVs have been observed between identical twins who otherwise have identical genomes.

Role in Disease

Like other types of genetic variation, some CNVs have been associated with susceptibility or resistance to disease. Gene copy number can be elevated in cancer cells. For instance, the EGFR copy number can be higher than normal in non-small cell lung cancer. In addition, a higher copy number of CCL3L1 has been associated with lower susceptibility to HIV infection, and a low copy number of FCGR3B (the CD16 cell surface immunoglobulin receptor) can increase susceptibility to systemic lupus erythematosus and similar inflammatory autoimmune disorders. Copy number variation has also been associated with autism, schizophrenia, and idiopathic learning disability. However, although once touted as the explanation for the elusive hereditary causes of complex diseases like rheumatoid arthritis, the most common CNVs have little or no role in causing disease.

Among common functional CNVs, gene gains outnumber losses, suggesting that many of them are favored in evolution and, therefore, beneficial in some way. One example of CNV is the human salivary amylase gene (AMY1). This gene is typically present as two diploid copies in chimpanzees. Humans average over 6 copies and may have as many as 15. This is thought to be an adaptation to a high-starch diet that improves the ability to digest starchy foods.

Gene Expression

Gene expression is the process by which information from a gene is used in the synthesis of a functional gene product. These products are often proteins, but in non-protein coding genes such as ribosomal RNA (rRNA), transfer RNA (tRNA) or small nuclear RNA (snRNA) genes, the product is a functional RNA. The process of gene expression is used by all known life - eukaryotes (including multicellular organisms), prokaryotes (bacteria and archaea), possibly induced by viruses - to generate the macromolecular machinery for life. Several steps in the gene expression process may be modulated, including the transcription, RNA splicing, translation, and post-translational modification of a protein. Gene regulation gives the cell control over structure and function, and is the basis for cellular differentiation, morphogenesis and the versatility and adaptability of any organism. Gene regulation may also serve as a substrate for evolutionary change, since control of the timing, location, and amount of gene expression can have a profound effect on the functions (actions) of the gene in a cell or in a multicellular organism. In genetics, gene expression is the most fundamental level at which the genotype gives rise to the phenotype. The genetic code stored in DNA is "interpreted" by gene expression, and the properties of the expression give rise to the organism's phenotype.

Mechanism

Transcription

Figure: *The process of transcription is carried out by RNA polymerase (RNAP), which uses DNA (black) as a template and produces RNA (blue).*

The gene itself is typically a long stretch of DNA which carries genetic information encoded by genetic code. Every molecule of DNA consists of two strands, each of them having 5' and 3' ends oriented in anti-parallel direction. The coding strand contains the genetic information while template strand (non-coding strand) serves as a blueprint for the production of RNA.

The production of RNA copies of the DNA is called transcription, and is performed by RNA polymerase, which adds one RNA nucleotide at a time to a growing RNA strand. This RNA is complementary to the template 3' '! 5' DNA strand, which is itself complementary to the coding 5' '! 3' DNA strand. Therefore, the resulting 5' '! 3' RNA strand is identical to the coding DNA strand with the exception that thymines (T) are replaced with uracils (U) in the RNA. A coding DNA strand reading "ATG" is transcribed as "AUG" in RNA.

Transcription in prokaryotes is carried out by a single type of RNA polymerase, which needs DNA sequence called Pribnow box and sigma factor (ó factor) to start transcription. In eukaryotes, the transcription is done by three types of RNA polymerases, each of them needs special DNA sequence called promoter and a set of DNA-binding proteins - transcription factors to initiate the process. RNA polymerase I is responsible for transcription of rRNA genes, while RNA polymerase II transcribes all protein-coding genes but also some non-coding RNAs (e.g. snRNAs, snoRNAs or long non-coding RNAs) as well. It contains special part called C-terminal domain (CTD) that is rich of serines, which after being phosphorylated accumulate factors necessary for RNA modification and maturation. RNA polymerase III transcribes 5S rRNA and tRNA genes but also some small non-coding RNA genes (e.g. 7SK). Transcription ends on a special sequence called terminator.

RNA Processing

While transcription of prokaryotic protein-coding genes creates messenger RNA (mRNA) which is ready for translation, transcription of eukaryotic genes leaves a primary transcript of RNA (pre-mRNA), which first has to undergo series of modification to become a mature mRNA.

These include 5' capping, which is set of enzymatic reactions that add 7-methylguanosine (m G) to the 5' end of pre-mRNA and thus protect the RNA from degradation by exonucleases. The m G cap is then bound by cap binding complex heterodimer (CBC20/CBC80) which aids in mRNA export to cytoplasm and also protect the RNA from decapping.

Another modification is 3' cleavage and polyadenylation. They occur if polyadenylation signal sequence (5'- AAUAAA-3') is present in pre-mRNA,which is usually between protein-coding sequence and terminator. The pre-mRNA is first cleaved and then a series of ~200 adenines (A) are added to form poly(A) tail which protects the RNA from degradation. Poly(A) tail is bound by multiple poly(A)-binding proteins (PABP) necessary for mRNA export and translation re-initiation. Very important modification of eukaryotic pre-mRNA is RNA splicing. Majority of eukaryotic pre-mRNAs consist of alternating segments called exons and introns. During the process of splicing, RNA-protein catalytical complex known as spliceosome, catalyze two transesterification reactions, which remove intron and release it in form of lariat structure and then splice neighbouring exons together. In certain cases, some introns or exons can be either removed or retained in mature mRNA. This so-called alternative splicing creates series of different transcripts originating from a single gene. Because these transcripts can be potentially translated into different proteins, splicing extends the complexity of eukaryotic gene expression.

Extensive RNA processing may be an evolutionary advantage made possible by the nucleus of eukaryotes. In prokaryotes transcription and translation happen together whilst in eukaryotes the nuclear membrane separates the two processes giving time for RNA processing to occur.

Non-coding RNA Maturation

In most organisms non-coding genes (ncRNA) are transcribed as precursors which undergo further processing. In the case of ribosomal RNAs (rRNA), they are often transcribed as a pre-rRNA which contains one or more rRNAs, the pre-rRNA is cleaved and modified (22 -O-methylation and pseudouridine formation) at a specific sites by approximately 150 different small nucleolus-restricted RNA species, called snoRNAs. SnoRNAs associate with proteins, forming snoRNPs. While snoRNA part basepair with the target RNA and thus position the modification to precise site, the protein part performs the catalytical reaction.

In eukaryotes, in particular a snoRNP, called RNase MRP cleaves the 45S pre-rRNA into the 28S, 5.8S, and 18S rRNAs. The rRNA and RNA processing factors form large aggregates called the nucleolus. In the case of transfer RNA (tRNA), for example, the 5' sequence is removed by RNase P, whereas the 3' end is removed by the tRNase Z enzyme and the non-templated 3' CCA tail is added by a nucleotidyl transferase.

In the case of micro RNA (miRNA), miRNAs are first transcribed as primary transcripts or pri-miRNA with a cap and poly-A tail and processed to short, 70-nucleotide stem-loop structures known as pre-miRNA in the cell nucleus by the enzymes Drosha and Pasha. After being exported, it is then processed to mature miRNAs in the cytoplasm by interaction with the endonuclease Dicer, which also initiates the formation of the RNA-induced silencing complex (RISC), composed of the Argonaute protein. Even snRNAs and snoRNAs themselves undergo series of modification before they become part of functional RNP complex. This is done either in the nucleoplasm or in the specialized compartments called Cajal bodies. Their bases are methylated or pseudouridinilated by a group of small Cajal body-specific RNAs (scaRNAs) which are structurally similar to snoRNAs.

RNA Export

In eukaryotes most mature RNA must be exported to the cytoplasm from the nucleus. While some RNAs function in the nucleus, many RNAs are transported through the nuclear pores and into the cytosol. Notably this includes all RNA types involved in protein synthesis. In some cases RNAs are additionally transported to a specific part of the cytoplasm, such as a synapse; they are then towed by motor proteins that bind through linker proteins to specific sequences (called "zipcodes") on the RNA.

Translation

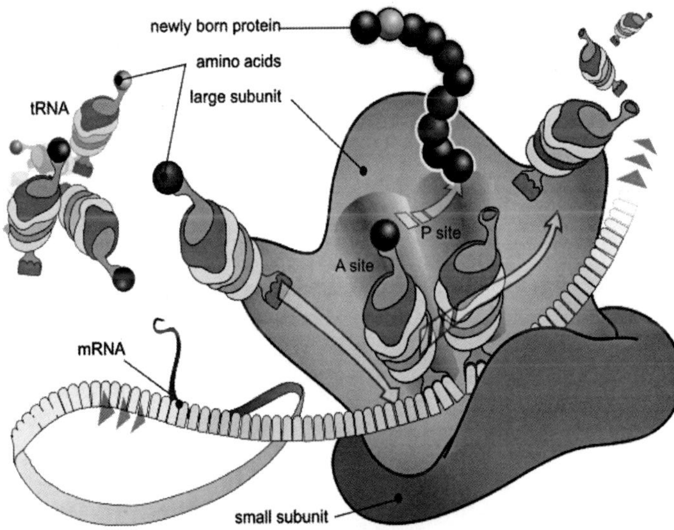

Figure: *During the translation, tRNA charged with amino acid enters the ribosome and aligns with the correct mRNA triplet. Ribosome then adds amino acid to growing protein chain.*

For some RNA (non-coding RNA) the mature RNA is the final gene product. In the case of messenger RNA (mRNA) the RNA is an information carrier coding for the synthesis of one or more proteins. mRNA carrying a single protein sequence (common in eukaryotes) is monocistronic whilst mRNA carrying multiple protein sequences (common in prokaryotes) is known as polycistronic. Every mRNA consists of three parts - 5' untranslated region (5'UTR), protein-coding region or open reading frame (ORF) and 3' untranslated region (3'UTR). Coding region carries information for protein synthesis encoded by genetic code into form of triplets. Each triplet of nucleotides of the coding region is called codon and corresponds to a binding site complementary to an anticodon triplet in transfer RNA. Transfer RNAs with the same anticodon sequence always carry identical type of amino acid. Amino acids are then chained together by the ribosome according to order of triplets in the coding region. The ribosome helps transfer RNA to bind to messenger RNA and takes the amino acid from each transfer RNA and makes a structure-less protein out of it. Each mRNA molecule is translated into many protein molecules, on average ~900 in mammals.

In prokaryotes translation generally occurs at the point of transcription (co-transcriptionally), often using a messenger RNA which is still in the process of being created. In eukaryotes translation can occur in a variety of regions of the cell depending on where the protein being written is supposed to be. Major locations are the cytoplasm for soluble cytoplasmic proteins and the membrane of endoplasmic reticulum for proteins which are for export from the cell or insertion into a cell membrane.

Proteins which are supposed to be expressed at the endoplasmic reticulum are recognised part-way through the translation process. This is governed by the signal recognition particle - a protein which binds to the ribosome and directs it to the endoplasmic reticulum when it finds a signal sequence on the growing (nascent) amino acid chain.

Folding

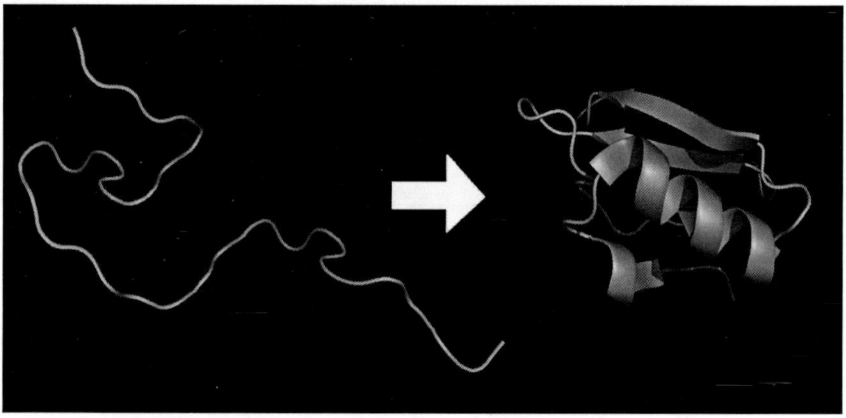

Figure: *Protein before (left) and after (right) folding.*

The polypeptide folds into its characteristic and functional three-dimensional structure from random coil. Each protein exists as an unfolded polypeptide or random coil when translated from a sequence of mRNA to a linear chain of amino acids.

This polypeptide lacks any developed three-dimensional structure (the left hand side of the neighboring figure). Amino acids interact with each other to produce a well-defined three-dimensional structure, the folded protein (the right hand side of the figure), known as the native state. The resulting three-dimensional structure is determined by the amino acid sequence (Anfinsen's dogma). The correct three-dimensional structure is essential to function, although some parts of functional proteins may remain unfolded Failure to fold into the intended shape usually produces inactive proteins with different properties including toxic prions. Several neurodegenerative and other diseases are believed to result from the accumulation of misfolded (incorrectly folded) proteins. Many allergies are caused by the folding of the proteins, for the immune system does not produce antibodies for certain protein structures. Enzymes called chaperones assist the newly formed protein to attain (fold into) the 3-dimensional structure it needs to function. Similarly, RNA chaperones help RNAs attain their functional shapes. Assisting protein folding is one of the main roles of the endoplasmic reticulum in eukaryotes.

Protein Transport

Many proteins are destined for other parts of the cell than the cytosol and a wide range of signalling sequences are used to direct proteins to where they are supposed to be. In prokaryotes this is normally a simple process due to limited compartmentalisation of the cell. However in eukaryotes there is a great variety of different targeting processes to ensure the protein arrives at the correct organelle. Not all proteins remain within the cell and many are exported, for example digestive enzymes, hormones and extracellular matrix proteins. In eukaryotes the export pathway is well developed and the main mechanism for the export of these proteins is translocation to the endoplasmic reticulum, followed by transport via the Golgi apparatus.

Regulation of Gene Expression

Figure: *The patchy colours of a tortoiseshell cat are the result of different levels of expression of pigmentation genes in different areas of the skin.*

Regulation of gene expression refers to the control of the amount and timing of appearance of the functional product of a gene. Control of expression is vital to allow a cell to produce the gene products it needs when it needs them; in turn this gives cells the flexibility to adapt to a variable environment, external signals, damage to the cell, etc. Some simple examples of where gene expression is important are:

- Control of insulin expression so it gives a signal for blood glucose regulation
- X chromosome inactivation in female mammals to prevent an "overdose" of the genes it contains.
- Cyclin expression levels control progression through the eukaryotic cell cycle

More generally gene regulation gives the cell control over all structure and function, and is the basis for cellular differentiation, morphogenesis and the versatility and adaptability of any organism. Any step of gene expression may be modulated, from the DNA-RNA transcription step to post-translational modification of a protein.

The stability of the final gene product, whether it is RNA or protein, also contributes to the expression level of the gene - an unstable product results in a low expression level.

In general gene expression is regulated through changes in the number and type of interactions between molecules that collectively influence transcription of DNA and translation of RNA.

Numerous terms are used to describe types of genes depending on how they are regulated, these include:

- A constitutive gene is a gene that is transcribed continually compared to a facultative gene which is only transcribed when needed.
- A housekeeping gene is typically a constitutive gene that is transcribed at a relatively constant level. The housekeeping gene's products are typically needed for maintenance of the cell. It is generally assumed that their expression is unaffected by experimental conditions. Examples include actin, GAPDH and ubiquitin.
- A facultative gene is a gene which is only transcribed when needed compared to a constitutive gene.
- An inducible gene is a gene whose expression is either responsive to environmental change or dependent on the position in the cell cycle.

Transcriptional Regulation

Regulation of transcription can be broken down into three main routes of influence; genetic (direct interaction of a control factor with the gene), modulation (interaction of a control factor with the transcription machinery) and epigenetic (non-sequence changes in DNA structure which influence transcription).

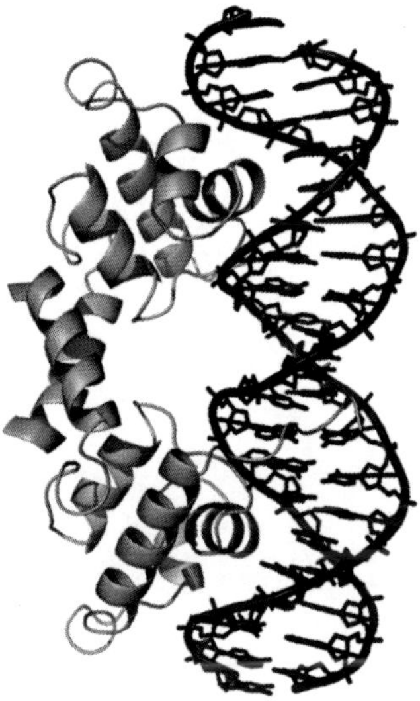

Figure: *The lambda repressor transcription factor (green) binds as a dimer to major groove of DNA target (red and blue) and disables initiation of transcription. From PDB 1LMB.*

Direct interaction with DNA is the simplest and the most direct method by which a protein can change transcription levels. Genes often have several protein binding sites around the coding region with the specific function of regulating transcription. There are many classes of regulatory DNA binding sites known as enhancers, insulators, repressors and silencers. The mechanisms for regulating transcription are very varied, from blocking key binding sites on the DNA for RNA polymerase to acting as an activator and promoting transcription by assisting RNA polymerase binding. The activity of transcription factors is further modulated by intracellular signals causing protein post-translational modification including phosphorylated, acetylated, or glycosylated. These changes influence a transcription factor's ability to bind, directly or indirectly, to promoter DNA, to recruit RNA polymerase, or to favor elongation of a newly synthetized RNA molecule. The nuclear membrane in eukaryotes allows further regulation of transcription factors by the duration of their presence in the nucleus which is regulated by reversible changes in their structure and by binding of other proteins. Environmental stimuli or endocrine signals may cause modification of regulatory proteins eliciting cascades of intracellular signals, which result in regulation of gene expression.

More recently it has become apparent that there is a huge influence of non-DNA-sequence specific effects on translation. These effects are referred to as epigenetic and involve the higher order structure of DNA, non-sequence specific DNA binding proteins and chemical modification of DNA. In general epigenetic effects alter the accessibility of DNA to proteins and so modulate transcription.

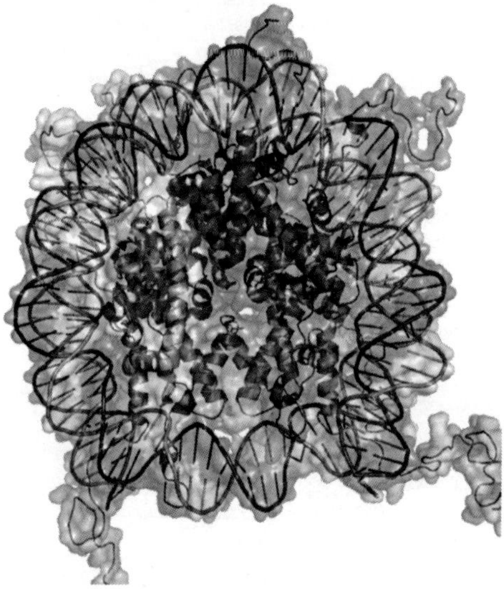

Figure: *In eukaryotes, DNA is organized in form of nucleosomes. Note how the DNA (blue and green) is tightly wrapped around the protein core made of histone octamer (ribbon coils), restricting access to the DNA. From PDB 1KX5.*

DNA methylation is a widespread mechanism for epigenetic influence on gene expression and is seen in bacteria and eukaryotes and has roles in heritable transcription silencing and transcription regulation. In eukaryotes the structure of chromatin, controlled by the histone code, regulates access to DNA with significant impacts on the expression of genes in euchromatin and heterochromatin areas.

Post-transcriptional Regulation

In eukaryotes, where export of RNA is required before translation is possible, nuclear export is thought to provide additional control over gene expression. All transport in and out of the nucleus is via the nuclear pore and transport is controlled by a wide range of importin and exportin proteins.

Expression of a gene coding for a protein is only possible if the messenger RNA carrying the code survives long enough to be translated. In a typical cell an RNA molecule is only stable if specifically protected from degradation. RNA degradation has particular importance in regulation of expression in eukaryotic cells where mRNA has to travel significant distances before being translated.

In eukaryotes RNA is stabilised by certain post-transcriptional modifications, particularly the 5' cap and poly-adenylated tail. Intentional degradation of mRNA is used not just as a defence mechanism from foreign RNA (normally from viruses) but also as a route of mRNA destabilisation. If an mRNA molecule has a complementary sequence to a small interfering RNA then it is targeted for destruction via the RNA interference pathway.

Techniques and Tools

The following experimental techniques are used to measure gene expression and are listed in roughly chronological order, starting with the older, more established technologies. They are divided into two groups based on their degree of multiplexity.

- Low-to-mid-plex techniques:
  - o Reporter gene

In molecular biology, a reporter gene (often simply reporter) is a gene that researchers attach to a regulatory sequence of another gene of interest in cell culture, animals or plants. Certain genes are chosen as reporters because the characteristics they confer on organisms expressing them are easily identified and measured, or because they are selectable markers. Reporter genes are often used as an indication of whether a certain gene has been taken up by or expressed in the cell or organism population.

Common Reporter Genes

To introduce a reporter gene into an organism, scientists place the reporter gene and the gene of interest in the same DNA construct to be inserted into the cell or organism. For bacteria or prokaryotic cells in culture, this is usually in the form of a circular DNA molecule called a plasmid. It is important to use a reporter gene that is not natively expressed in the cell or organism under study, since the expression of the reporter is being used as a marker for successful uptake of the gene of interest. Commonly used reporter genes that induce visually identifiable characteristics usually involve fluorescent and luminescent proteins. Examples include the gene that encodes jellyfish green fluorescent protein (GFP), which causes cells that express it to glow green under blue light, the enzyme luciferase, which catalyzes a reaction with luciferin to produce light, and the red fluorescent protein from the gene dsRed. The GUS gene has been commonly used in plants but luciferase and GFP are becoming more common.

A common reporter in bacteria is the LacZ gene, which encodes the protein beta-galactosidase. This enzyme causes bacteria expressing the gene to appear blue when grown on a medium that contains the substrate analog X-gal. An example of a selectable-marker which is also a reporter in bacteria is the chloramphenicol acetyltransferase (CAT) gene, which confers resistance to the antibiotic chloramphenicol.

2

Transformation and Transfection of Gene

Many methods of transfection and transformation - two ways of expressing a foreign or modified gene in an organism - are effective in only a small percentage of a population subjected to the techniques. Thus, a method for identifying those few successful gene uptake events is necessary. Reporter genes used in this way are normally expressed under their own promoter independent from that of the introduced gene of interest; the reporter gene can be expressed constitutively (that is, it is "always on") or inducibly with an external intervention such as the introduction of IPTG in the â-galactosidase system.

As a result, the reporter gene's expression is independent of the gene of interest's expression, which is an advantage when the gene of interest is only expressed under certain specific conditions or in tissues that are difficult to access. In the case of selectable-marker reporters such as CAT, the transfected population of bacteria can be grown on a substrate that contains chloramphenicol. Only those cells that have successfully taken up the construct containing the CAT gene will survive and multiply under these conditions.

Gene Expression Assays

Reporter genes can also be used to assay for the expression of the gene of interest, which may produce a protein that has little obvious or immediate effect on the cell culture or organism. In these cases the reporter is directly attached to the gene of interest to create a gene fusion. The two genes are under the same promoter elements and are transcribed into a single messenger RNA molecule. The mRNA is then translated into protein. In these cases it is important that both proteins be able to properly fold into their active conformations and interact with their substrates despite being fused. In building the DNA construct, a segment of DNA coding for a flexible polypeptide linker region is usually included so that the reporter and the gene product will only minimally interfere with one another.

Promoter Assays

Reporter genes can be used to assay for the activity of a particular promoter in a cell or organism. In this case there is no separate "gene of interest"; the reporter gene is simply placed under the control of the target promoter and the reporter gene product's activity is quantitatively measured. The results are normally reported relative to the activity under a "consensus" promoter known to induce strong gene expression.

Further Uses

A more complex usage of reporter genes on a large scale is in two-hybrid screening, which aims to identify proteins that natively interact with one another in vivo.

GUS Reporter System

The GUS reporter system (GUS: beta-glucuronidase) is a reporter gene system, particularly useful in plant molecular biology. Several kinds of GUS reporter gene assay are actually available, depending on the substrate used. The term GUS staining refers to the most common of these, a histochemical technique.

Purpose

The purpose of this technique is to analyze the activity of a promoter (in terms of expression of a gene under that promoter) either in a quantitative way or through visualization of its activity in different tissues.

The technique is based on beta-glucuronidase, an enzyme from the bacterium Escherichia coli; this enzyme, when incubated with some specific colorless or non-fluorescent substrates, can transform them into coloured or fluorescent products.

Substrates

There are actually different possible glucuronides that can be used as substrates for the beta-glucuronidase, depending on the type of detection needed (histochemical, spectrophotometrical, fluorimetrical).

The most common substrate for GUS histochemical staining is 5-bromo-4-chloro-3-indolyl glucuronide (X-Gluc): the product of the reaction is in this case a clear blue color. Other common substrates are p-nitrophenyl â-D-glucuronide for the spectrophotometrical assay and 4-methylumbelliferyl-beta-D-glucuronide (MUG) for the fluorimetrical assay.

History

The system was originally developed by Richard Anthony Jefferson during his Ph.D. at the University of Colorado at Boulder. He adapted the technique for the use with plants as he worked in the Plant Breeding Institute of Cambridge, between 1985 and 1987 . Since then thousands of labs have used the system, making it probably the most widely used tool in plant molecular biology, as underlined by over 6000 citations in scientific literature.

Target Organisms

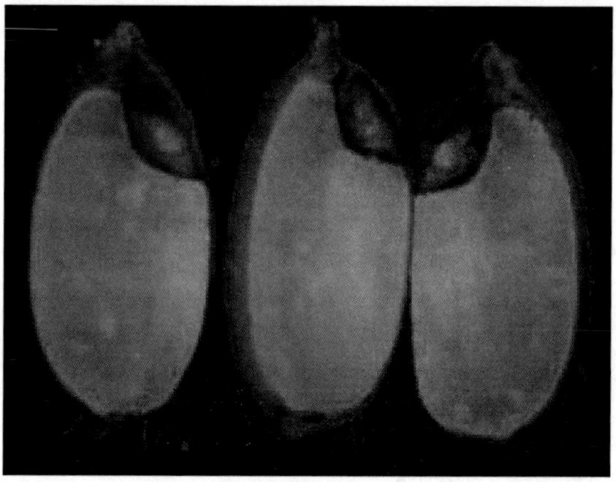

Figure

Rice Embryo Showing GUS Expression

An organism is suitable for a GUS assay if it has no beta-glucuronidase or if the activity is very low (background activity). For this reason the assay is not useful in most vertebrates and many molluscs. Since there is no detectable GUS activity in higher plants, mosses, algae, ferns, fungi and most bacteria, the assay is perfectly suited for these organisms. Thus it is used widely in plant science.

Other Reporter Systems

The GUS system is not the only available gene reporter system for the analysis of promoter activity. Other competing systems are based on e.g. luciferase, GFP, beta-galactosidase, chloramphenicol acetyltransferase (CAT), alkaline phosphatase. The use of one or the other system is mainly dependent on the organism of interest.

Other Uses

The GUS assay, as well as other reporter gene systems, can be used for other kinds of studies other than the classical promoter activity assay. Reporter systems have been used for the determination of the efficiency of gene delivery systems, the intracellular localization of a gene product, the detection of protein-protein or protein-DNA interactions, the efficiency of translation initiation signals and the success of molecular cloning efforts.

o Northern blot

The northern blot is a technique used in molecular biology research to study gene expression by detection of RNA (or isolated mRNA) in a sample.

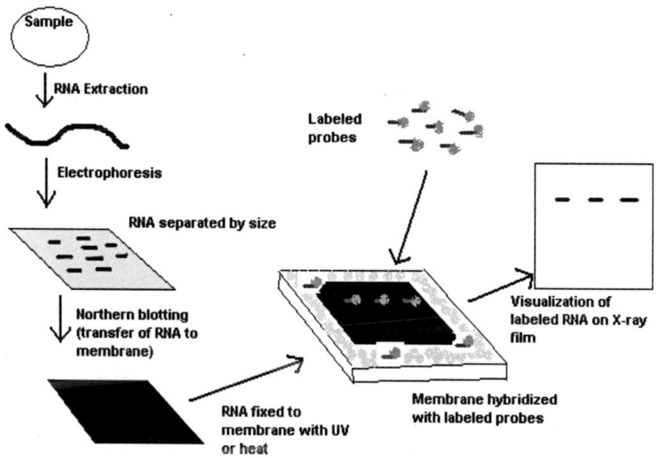

Figure: *Flow diagram outlining the general procedure for RNA detection by northern blotting.*

With northern blotting it is possible to observe cellular control over structure and function by determining the particular gene expression levels during differentiation, morphogenesis, as well as abnormal or diseased conditions. Northern blotting involves the use of electrophoresis to separate RNA samples by size and detection with a hybridization probe complementary to part of or the entire target sequence. The term 'northern blot' actually refers specifically to the capillary transfer of RNA from the electrophoresis gel to the blotting membrane. However, the entire process is commonly referred to as northern blotting. The northern blot technique was developed in 1977 by James Alwine, David Kemp, and George Stark at Stanford University. Northern blotting takes its name from its similarity to the first blotting technique, the Southern blot, named for biologist Edwin Southern. The major difference is that RNA, rather than DNA, is analyzed in the northern blot.

Procedure

A general blotting procedure starts with extraction of total RNA from a homogenized tissue sample or from cells. Eukaryotic mRNA can then be isolated through the use of oligo (dT) cellulose chromatography to isolate only those RNAs with a poly(A) tail. RNA samples are then separated by gel electrophoresis. Since the gels are fragile and the probes are unable to enter the matrix, the RNA samples, now separated by size, are transferred to a nylon membrane through a capillary or vacuum blotting system.

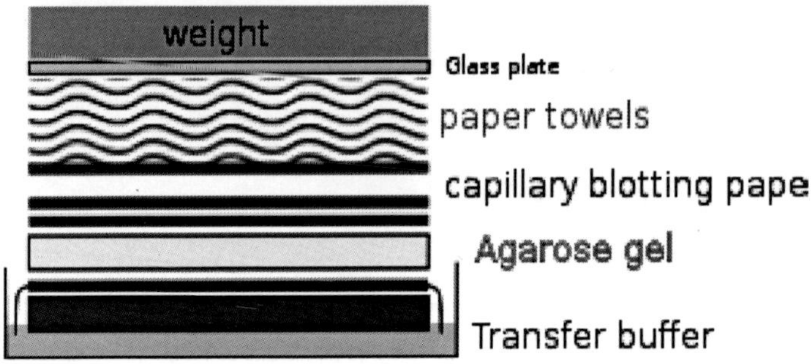

Figure: *Capillary blotting system setup for the transfer of RNA from an electrophoresis gel to a blotting membrane.*

A nylon membrane with a positive charge is the most effective for use in northern blotting since the negatively charged nucleic acids have a high affinity for them. The transfer buffer used for the blotting usually contains formamide because it lowers the annealing temperature of the probe-RNA interaction, thus preventing RNA degradation by high temperatures. Once the RNA has been transferred to the membrane, it is immobilized through covalent linkage to the membrane by UV light or heat.

After a probe has been labeled, it is hybridized to the RNA on the membrane. Experimental conditions that can affect the efficiency and specificity of hybridization include ionic strength, viscosity, duplex length, mismatched base pairs, and base composition. The membrane is washed to ensure that the probe has bound specifically and to avoid background signals from arising. The hybrid signals are then detected by X-ray film and can be quantified by densitometry. To create controls for comparison in a northern blot, samples not displaying the gene product of interest can be used after determination by microarrays or RT-PCR.

Gels

The RNA samples are most commonly separated on agarose gels containing formaldehyde as a denaturing agent for the RNA to limit secondary structure. The gels can be stained with ethidium bromide (EtBr) and viewed under UV light to observe thc quality and quantity of RNA before blotting. Polyacrylamide gel electrophoeresis with urea can also be used in RNA separation but it is most commonly used for fragmented RNA or microRNAs. An RNA ladder is often run alongside the samples on an electrophoresis gel to observe the size of fragments obtained but in total RNA samples the ribosomal subunits can act as size markers. Since the large ribosomal subunit is 28S (approximately

5kb) and the small ribosomal subunit is 18S (approximately 2kb) two prominent bands will appear on the gel, the larger at close to twice the intensity of the smaller.

Probes

Probes for northern blotting are composed of nucleic acids with a complementary sequence to all or part of the RNA of interest, they can be DNA, RNA, or oligonucleotides with a minimum of 25 complementary bases to the target sequence. RNA probes (riboprobes) that are transcribed in vitro are able to withstand more rigorous washing steps preventing some of the background noise. Commonly cDNA is created with labelled primers for the RNA sequence of interest to act as the probe in the northern blot.

The probes need to be labelled either with radioactive isotopes (P) or with chemiluminescence in which alkaline phosphatase or horseradish peroxidase breakdown chemiluminescent substrates producing a detectable emission of light. The chemiluminescent labelling can occur in two ways: either the probe is attached to the enzyme, or the probe is labelled with a ligand (e.g. biotin) for mwhich the antibody (e.g. avidin or streptavidin) is attached to the enzyme. X-ray film can detect both the radioactive and chemiluminescent signals and many researchers prefer the chemiluminescent signals because they are faster, more sensitive, and reduce the health hazards that go along with radioactive labels. The same membrane can be probed up to five times without a significant loss of the target RNA.

Applications

Northern blotting allows one to observe a particular gene's expression pattern between tissues, organs, developmental stages, environmental stress levels, pathogen infection, and over the course of treatment. The technique has been used to show overexpression of oncogenes and downregulation of tumor-suppressor genes in cancerous cells when compared to 'normal' tissue, as well as the gene expression in the rejection of transplanted organs. If an upregulated gene is observed by an abundance of mRNA on the northern blot the sample can then be sequenced to determine if the gene is known to researchers or if it is a novel finding.

The expression patterns obtained under given conditions can provide insight into the function of that gene. Since the RNA is first separated by size, if only one probe type is used variance in the level of each band on the membrane can provide insight into the size of the product, suggesting alternative splice products of the same gene or repetitive sequence motifs. The variance in size of a gene product can also indicate deletions or errors in transcript processing, by altering the probe target used along the known sequence it is possible to determine which region of the RNA is missing.

BlotBase is an online database publishing northern blots. BlotBase has over 700 published northern blots of human and mouse samples, in over 650 genes across more than 25 different tissue types. Northern blots can be searched by a blot ID, paper reference, gene identifier, or by tissue. The results of a search provide the blot ID, species, tissue, gene, expression level, blot image (if available), and links to the publication that the work originated from. This new database provides sharing of information between members of the science community that was not previously seen in northern blotting as it was in sequence analysis, genome determination, protein structure, etc.

Advantages and Disadvantages

Analysis of gene expression can be done by several different methods including RT-PCR, RNase protection assays, microarrays, serial analysis of gene expression (SAGE), as well as northern blotting. Microarrays are quite commonly used and are usually consistent with data obtained from northern blots; however, at times northern blotting is able to detect small changes in gene expression that microarrays cannot. The advantage that microarrays have over northern blots is that thousands of genes can be visualized at a time, while northern blotting is usually looking at one or a small number of genes.

A problem in northern blotting is often sample degradation by RNases (both endogenous to the sample and through environmental contamination), which can be avoided by proper sterilization of glassware and the use of RNase inhibitors such as DEPC (diethylpyrocarbonate). The chemicals used in most northern blots can be a risk to the researcher, since formaldehyde, radioactive material, ethidium bromide, DEPC, and UV light are all harmful under certain exposures. Compared to RT-PCR, northern blotting has a low sensitivity, but it also has a high specificity which is important to reduce false positive results.

The advantages of using northern blotting include the detection of RNA size, the observation of alternate splice products, the use of probes with partial homology, the quality and quantity of RNA can be measured on the gel prior to blotting, and the membranes can be stored and reprobed for years after blotting.

Reverse Northern Blot

A variant of the procedure known as the reverse northern blot is occasionally used. In this procedure, the substrate nucleic acid (that is affixed to the membrane) is a collection of isolated DNA fragments, and the probe is RNA extracted from a tissue and radioactively labelled.

The use of DNA microarrays that have come into widespread use in the late 1990s and early 2000s is more akin to the reverse procedure, in that they involve the use of isolated DNA fragments affixed to a substrate, and hybridization with a probe made from cellular RNA. Thus the reverse procedure, though originally uncommon, enabled northern analysis to evolve into gene expression profiling, in which many (possibly all) of the genes in an organism may have their expression monitored.

Fluorescent in Situ Hybridization

FISH (fluorescence in situ hybridization) is a cytogenetic technique developed by Christoph Lengauer that is used to detect and localize the presence or absence of specific DNA sequences on chromosomes. FISH uses fluorescent probes that bind to only those parts of the chromosome with which they show a high degree of sequence complementarity. Fluorescence microscopy can be used to find out where the fluorescent probe bound to the chromosomes. FISH is often used for finding specific features in DNA for use in genetic counselling, medicine, and species identification. FISH can also be used to detect and localize specific mRNAs within tissue samples. In this context, it can help define the spatial-temporal patterns of gene expression within cells and tissues.

Probes

Probes are often derived from fragments of DNA that were isolated, purified, and amplified for use in the Human Genome Project. The size of the human genome is so large, compared to the length that could be sequenced directly, that it was necessary to divide the genome into fragments. (In the eventual analysis, these fragments were put into order by digesting a copy of each fragment into still smaller fragments using sequence-specific endonucleases, measuring the size of each small fragment using size-exclusion chromatography, and using that information to determine where the large fragments overlapped one another.)

To preserve the fragments with their individual DNA sequences, the fragments were added into a system of continually replicating bacteria populations. Clonal populations of bacteria, each population maintaining a single artificial chromosome, are stored in various laboratories around the world. The artificial chromosomes (BAC) can be grown, extracted, and labeled, in any lab. These fragments are on the order of 100 thousand base-pairs, and are the basis for most FISH probes.

Preparation and Hybridization Process

First, a probe is constructed. The probe must be large enough to hybridize specifically with its target but not so large as to impede the hybridization process. The probe is tagged directly with fluorophores, with targets for antibodies or with biotin. Tagging can be done in various ways, such as nick translation, or PCR using tagged nucleotides. Then, an interphase or metaphase chromosome preparation is produced. The chromosomes are firmly attached to a substrate, usually glass. Repetitive DNA sequences must be blocked by adding short fragments of DNA to the sample. The probe is then applied to the chromosome DNA and incubated for approximately 12 hours while hybridizing. Several wash steps remove all unhybridized or partially-hybridized probes.

The results are then visualized and quantified using a microscope that is capable of exciting the dye and recording images. If the fluorescent signal is weak, amplification of the signal may be necessary in order to exceed the detection threshold of the microscope. Fluorescent signal strength depends on many factors such as probe labeling efficiency, the type of probe, and the type of dye.

The dye incorporation rate can be quantified photometric with specialized nano-volume photometer (starting with 0.3 μl probe volume). Fluorescently-tagged antibodies or streptavidin are bound to the dye molecule. These secondary components are selected so that they have a strong signal.

FISH experiments designed to detect or localize gene expression within cells and tissues rely on the use of a reporter gene, such as one expressing green fluorescent protein, to provide the fluorescence signal.

Variations on Probes and Analysis

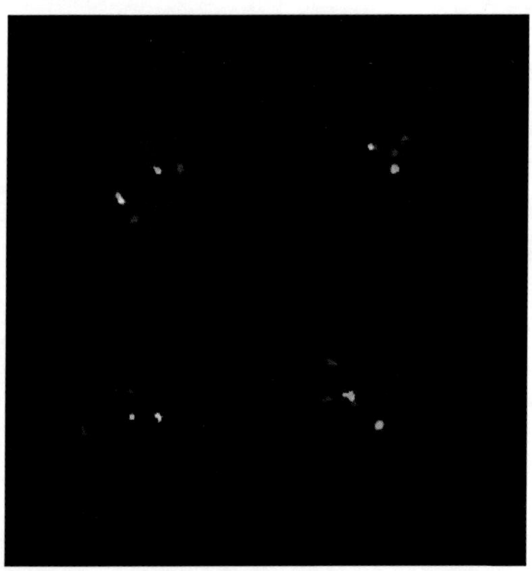

Figure: *Interphase cells positive for a chromosomal t(9;22) rearrangement.*

FISH is a very general technique. The differences between the various FISH techniques are usually due to variations in the sequence and labeling of the probes; and how they are used in combination. These few modifications make possible all FISH techniques.

Probe size is important because longer probes hybridize less specifically than shorter probes. The overlap defines the resolution of detectable features. For example, if the goal of an experiment is to detect the breakpoint of a translocation, then the overlap of the probes — the degree to which one DNA sequence is contained in the adjacent probes — defines the minimum window in which the breakpoint may be detected. The mixture of probe sequences determines the type of feature the probe can detect. Probes that hybridize along an entire chromosome are used to count the number of a certain chromosome, show translocations, or identify extra-chromosomal fragments of chromatin. This is often called "whole-chromosome painting." If every possible probe is used, every chromosome, (the whole genome) would be marked fluorescently, which would not be particularly useful for determining features of individual sequences. However, a mixture of smaller probes can be created that is specific to a particular region (locus) of DNA; these mixtures are used to detect deletion mutations. When combined with a specific colour, a locus-specific probe mixture is used to detect very specific translocations. Special locus-specific probe mixtures are often used to count chromosomes, by binding to the centromeric regions of chromosomes, which are unique enough to identify each chromosome (with the exception of Chromosome 13, 14, 21, 22.)

A variety of other techniques use mixtures of differently-colored probes. A range of colors in mixtures of fluorescent dyes can be detected, so each human chromosome can be identified by a characteristic color using whole-chromosome probe mixtures and a variety of ratios of colors. Although there are more chromosomes than easily-distinguishable fluorescent dye colors, ratios of probe mixtures can be used to create secondary colors. Similar to comparative genomic hybridization, the probe mixture for the secondary colors is created by mixing the correct ratio of two sets of differently-colored probes for the same chromosome. This technique is sometimes called M-FISH. The same physics that make a variety of colors possible for M-FISH can be used for the detection of translocations. That is, colors that are adjacent appear to overlap; a secondary color is observed. Some assays are designed so that the secondary color will be present or absent in cases of interest.

An example is the detection of BCR/ABL translocations, where the secondary color indicates disease. This variation is often called double-fusion FISH or D-FISH. In the opposite situation—where the absence of the secondary color is pathological—is illustrated by an assay used to investigate translocations where only one of the breakpoints is known or constant. Locus-specific probes are made for one side of the breakpoint and the other intact chromosome. In normal cells, the secondary colour is observed, but only the primary colour is observed when the translocation occurs. This technique is sometimes called "break-apart FISH".

Stellaris FISH Probes

Stellaris FISH, formerly known as Single Molecule RNA FISH, is a method of detecting and quantifying mRNA and other long RNA molecules in a thin layer of tissue sample. Targets can be reliably imaged through the application of multiple short singly labeled oligonucleotide probes.

The binding of up to 48 fluorescent labeled oligos to a single molecule of mRNA provides sufficient fluorescence to accurately detect and localize each target mRNA in a wide-field fluorescent microscopy image. Probes not binding to the intended sequence do not achieve sufficient localized fluorescence to be distinguished from background. Single molecule RNA FISH assays can be performed in simplex or multiplex, and can be used as a follow-up experiment to qPCR, or imaged simultaneously with a fluorescent antibody assay. The technology has potential applications in cancer diagnosis, neuroscience, gene expression analysis, and companion diagnostics.

Fiber FISH

In an alternative technique to interphase or metaphase preparations, fiber FISH, interphase chromosomes are attached to a slide in such a way that they are stretched out in a straight line, rather than being tightly coiled, as in conventional FISH, or adopting a random conformation, as in interphase FISH. This is accomplished by applying mechanical shear along the length of the slide, either to cells that have been fixed to the slide and then lysed, or to a solution of purified DNA.

A technique known as chromosome combing is increasingly used for this purpose. The extended conformation of the chromosomes allows dramatically higher resolution - even down to a few kilobases. The preparation of fiber FISH samples, although conceptually simple, is a rather skilled art, and only specialized laboratories use the technique routinely.

Q-FISH

Q-FISH combines FISH with PNAs and computer software to quantify fluorescence intensity. This technique is used routinely in telomere length research.

Flow-FISH

Flow-FISH uses flow cytometry to perform FISH automatically using per-cell fluorescence measurements.

Medical Applications

Often parents of children with a developmental delay want to know more about their child's conditions before choosing to have another child. These concerns can be addressed by analysis of the parents' and child's DNA. In cases where the child's developmental delay is not understood, the cause of it can potentially be determined using FISH and cytogenetic techniques.

Examples of diseases that are diagnosed using FISH include Prader-Willi syndrome, Angelman syndrome, 22q13 deletion syndrome, chronic myelogenous leukemia, acute lymphoblastic leukemia, Cri-du-chat, Velocardiofacial syndrome, and Down syndrome. FISH on sperm cells is indicated for men with an abnormal somatic or meiotic karyotype as well as those with oligozoospermia, since approximately 50% of oligozoospermic men have an increased rate of sperm chromosome abnormalities. The analysis of chromosomes 21, X, and Y is enough to identify oligozoospermic individuals at risk.

In medicine, FISH can be used to form a diagnosis, to evaluate prognosis, or to evaluate remission of a disease, such as cancer. Treatment can then be specifically tailored.

A traditional exam involving metaphase chromosome analysis is often unable to identify features that distinguish one disease from another, due to subtle chromosomal features; FISH can elucidate these differences.

FISH can also be used to detect diseased cells more easily than standard Cytogenetic methods, which require dividing cells and requires labor and time-intensive manual preparation and analysis of the slides by a technologist. FISH, on the other hand, does not require living cells and can be quantified automatically, a computer counts the fluorescent dots present. However, a trained technologist is required to distinguish subtle differences in banding patterns on bent and twisted metaphase chromosomes.

Species Identification

FISH is often used in clinical studies. If a patient is infected with a suspected pathogen, bacteria, from the patient's tissues or fluids, are typically grown on agar to determine the identity of the pathogen. Many bacteria, however, even well-known species, do not grow well under laboratory conditions.

FISH can be used to detect directly the presence of the suspect on small samples of patient's tissue.

FISH can also be used to compare the genomes of two biological species, to deduce evolutionary relationships. A similar hybridization technique is called a zoo blot. Bacterial FISH probes are often primers for the 16s rRNA region.

FISH is widely used in the field of microbial ecology, to identify microorganisms. Biofilms, for example, are composed of complex (often) multi-species bacterial organizations.

Preparing DNA probes for one species and performing FISH with this probe allows one to visualize the distribution of this specific species within the biofilm. Preparing probes (in two different colors) for two species allows to visualize/study co-localization of these two species in the biofilm, and can be useful in determining the fine architecture of the biofilm.

Lab-on-a-chip and FISH

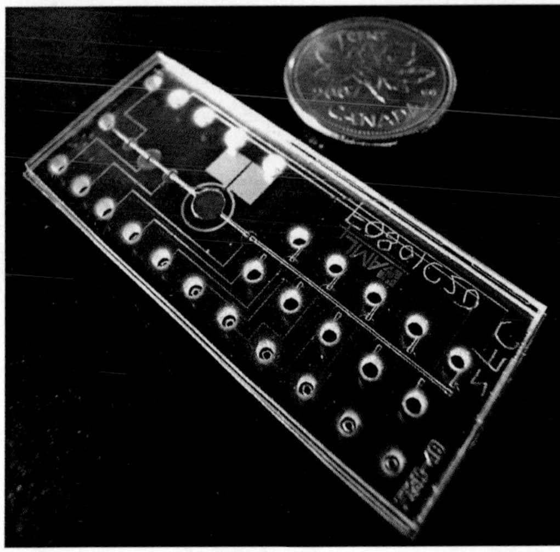

Figure

Microfluidic chip that automates the interphase FISH procedure. The microchip shown requires only minutes of setup time by technician, as opposed to the hours or days of labour needed to perform FISH with conventional equipment. Although interphase fluorescence in situ hybridization (FISH) is a sensitive diagnostic tool used for the detection of chromosomal abnormalities on cell-by-cell basis, the cost-per-test and the technical complexity of current FISH protocols has inhibited its widespread utilization. Lab-on-a-chip or microfluidic devices, incorporate networks of microchannels that can miniaturize, integrate and automate conventional analytical techniques onto chip-style platforms. Since microchannels permit sophisticated levels of fluid control (down to picolitres), these devices can reduce analysis times, lower reagent consumption, and minimize human intervention.

Currently, FISH has been performed on glass microfluidic platforms that standardize much of the protocol offering repeatable results that are accurate, cost-effective and easier to obtain in a clinical setting.

Compared to conventional FISH methods, these first implementations of on-chip FISH provide a 10-fold higher throughput and a 10-fold reduction in the cost of testing, enabling the simultaneous assessment of several chromosomal abnormalities or patients. It is increasingly essential that diagnostic tests determine the type and extent of chromosomal abnormalities for more informed diagnosis and for appropriate choice of treatment strategies. Since the on-chip FISH technique is 10-20 times more cost-effective than conventional methods, and can be fully integrated and automated, this technology will make widespread genetic testing of patients more accessible in a clinical setting. Recently, the first demonstration of Metaphase FISH on chip has led to renewed efforts towards automating the metaphase FISH protocol. Metaphase FISH had continued to be difficult to integrate owing to the complex sample preparation protocol often spanning over 3 weeks. New reports confirm that a research group in Denmark have tested successfully a novel lab on chip device to integrate the entire sample preparation protocol for Metaphase FISH called FISHprep.

Virtual Karyotype

Virtual karyotyping is another cost-effective, clinically available alternative to FISH panels uses thousands to millions of probes on a single array to detect copy number changes, genome-wide, at unprecedented resolution. Currently, this type of analysis will only detect gains and losses of chromosomal material and will not detect balanced rearrangements, such as translocations and inversions which are hallmark aberrations seen in many types of leukemia and lymphoma.

Molecular Cytogenetics

Molecular cytogenetics involves the combination of molecular biology and cytogenetics. In general this involves the use of a series of techniques referred to as fluorescence in situ hybridization, or FISH, in which DNA probes are labeled with different colored fluorescent tags to visualize one or more specific regions of the genome. FISH can either be performed as a direct approach to metaphase chromosomes or interphase nuclei. Alternatively, an indirect approach can be taken in which the entire genome can be assessed for copy number changes using virtual karyotyping. Virtual karyotypes are generated from arrays made of thousands to millions of probes, and computational tools are used to recreate the genome in silico.

Virtual Karyotype

Virtual karyotype detects genomic copy number variations at a higher resolution level than conventional karyotyping or chromosome-based comparative genomic hybridization (CGH).

Background

A karyotype (Fig 1) is the characteristic chromosome complement of a eukaryote species. A karyotype is typically presented as an image of the chromosomes from a single cell arranged from largest (chromosome 1) to smallest (chromosome 22), with the sex chromosomes (X and/or Y) shown last. Historically, karyotypes have been obtained by staining cells after they have been chemically arrested during cell division. Karyotypes have been used for several decades to identify chromosomal abnormalities in both germline and cancer cells. Conventional karyotypes can assess the entire genome for changes in chromosome structure and number, but the resolution is relatively coarse, with a detection limit of 5-10Mb.

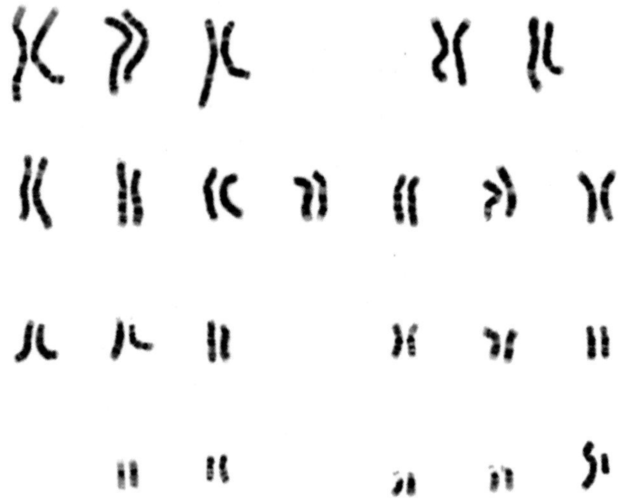

Figure: *Karyotype of human male using Giemsa staining*

Virtual Karyotype

Recently, platforms for generating high-resolution karyotypes in silico from disrupted DNA have emerged, such as array comparative genomic hybridization (arrayCGH) and SNP arrays. Conceptually, the arrays are composed of hundreds to millions of probes which are complementary to a region of interest in the genome. The disrupted DNA from the test sample is fragmented, labeled, and hybridized to the array. The hybridization signal intensities for each probe are used by specialized software to generate a log2ratio of test/normal for each probe on the array. Knowing the address of each probe on the array and the address of each probe in the genome, the software lines up the probes in chromosomal order and reconstructs the genome in silico (Fig 2 and 3).

Virtual karyotypes have dramatically higher resolution than conventional cytogenetics. The actual resolution will depend on the density of probes on the array. Currently, the Affymetrix SNP6.0 is the highest density commercially available array for virtual karyotyping applications. It contains 1.8 million polymorphic and non-polymorphic markers for a practical resolution of 10-20kb—about the size of a gene. This is approximately 1000-fold greater resolution than karyotypes obtained from conventional cytogenetics.

How to Get a Virtual Karyotype

Virtual karyotypes can be performed on germline samples for constitutional disorders, and clinical testing is available from dozens of CLIA certified laboratories (GeneTests.org).

Virtual karyotyping can also be done on fresh or formalin-fixed paraffin-embedded tumors. CLIA-certified laboratories offering testing on tumors include Creighton Medical Laboratories (fresh and paraffin embedded tumor samples) and CombiMatrix Molecular Diagnostics (fresh tumor samples).

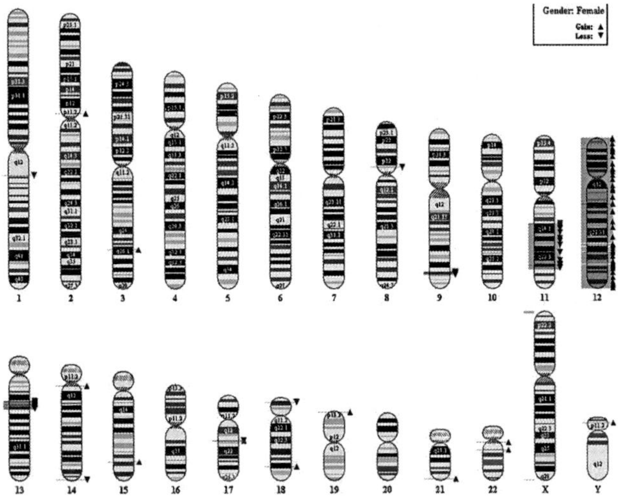

Figure: *Virtual karyotype of a chronic lymphocytic leukemia sample using a SNP array.*

Different Platforms for Virtual Karyotyping

Array-based karyotyping can be done with several different platforms, both laboratory-developed and commercial. The arrays themselves can be genome-wide (probes distributed over the entire genome) or targeted (probes for genomic regions known to be involved in a specific disease) or a combination of both. Further, arrays used for karyotyping may use non-polymorphic probes, polymorphic probes (i.e., SNP-containing), or a combination of both. Non-polymorphic probes can provide only copy number information, while SNP arrays can provide both copy number and loss-of-heterozygosity (LOH) status in one assay.

The probe types used for non-polymorphic arrays include cDNA, BAC clones (e.g., BlueGnome), and oligonucleotides (e.g., Agilent, Santa Clara, CA, USA or Nimblegen, Madison, WI, USA). Commercially available oligonucleotide SNP arrays can be solid phase (Affymetrix, Santa Clara, CA, USA) or bead-based (Illumina, SanDiego, CA, USA). Despite the diversity of platforms, ultimately they all use genomic DNA from disrupted cells to recreate a high resolution karyotype in silico. The end product does not yet have a consistent name, and has been called virtual karyotyping, digital karyotyping, molecular allelokaryotyping, and molecular karyotyping. Other terms used to describe the arrays used for karyotyping include SOMA (SNP oligonucleotide microarrays) and CMA (chromosome microarray). Some consider all platforms to be a type of array comparative genomic hybridization (arrayCGH), while others reserve that term for two-dye methods, and still others segregate SNP arrays because they generate more and different information than two-dye arrayCGH methods.

Applications

Detecting Copy-number Changes

Copy number changes can be seen in both germline and tumor samples. Copy number changes can be detected by arrays with non-polymorphic probes, such as arrayCGH, and by SNP-based arrays. Human beings are diploid, so a normal copy number is always two for the non-sex chromosomes. Deletions: A deletion is the loss of genetic material. The deletion can be heterozygous (copy number of 1) or homozygous (copy number of 0, nullisomy). Microdeletion syndromes are examples of constitutional disorders due to small deletions in germline DNA. Deletions in tumor cells may represent the inactivation of a tumor suppressor gene, and may have diagnostic, prognostic, or therapeutic implications.

Gains: A copy number gain represents the gain of genetic material. If the gain is of just one additional copy of a segment of DNA, it may be called a duplication. If there is one extra copy of an entire chromosome, it may be called a trisomy. Copy number gains in germline samples may be disease-associated or may be a benign copy number variant. When seen in tumor cells, they may have diagnostic, prognostic, or therapeutic implications.

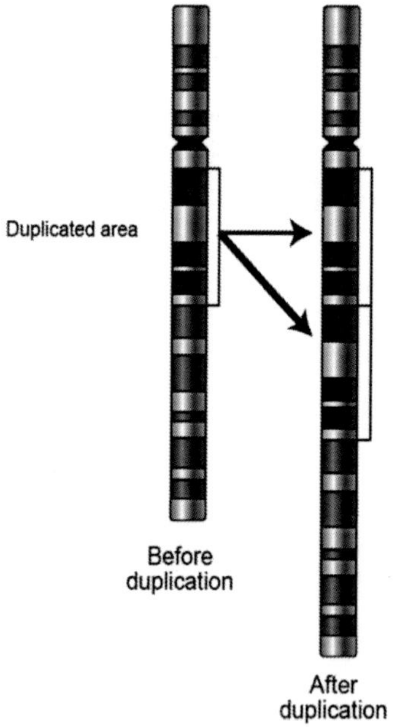

Figure: *Schematic of a region of a chromosome before & after a duplication event*

Amplifications: Technically, an amplification is a type of copy number gain in which there is a copy number >10. In the context of cancer biology, amplifications are often seen in oncogenes. This could indicate a worse prognosis, help categorize the tumor, or indicate drug eligibility. An example of drug eligibility is Her2Neu amplification and Herceptin, and an image of Her2Neu amplification detected by SNP array virtual karyotyping is provided.

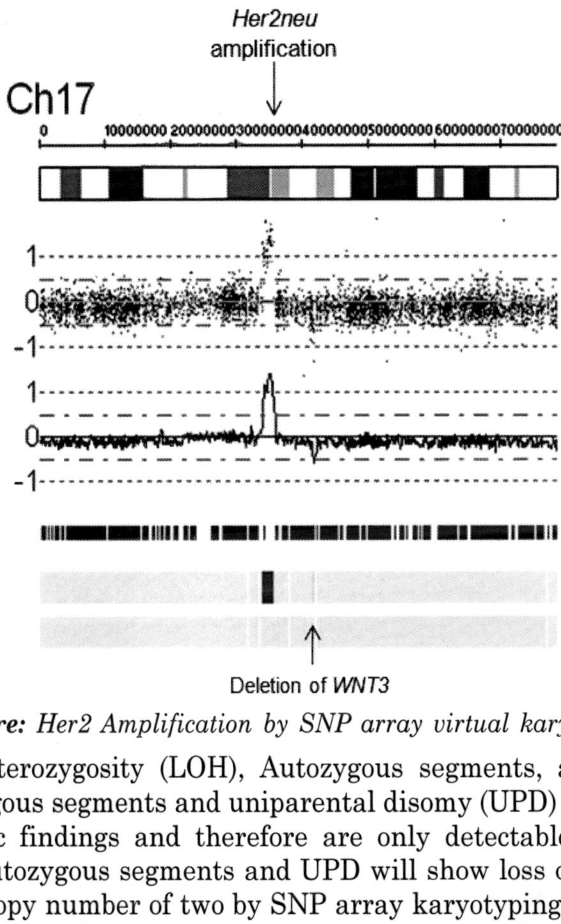

Figure: *Her2 Amplification by SNP array virtual karyotype.*

Loss of heterozygosity (LOH), Autozygous segments, and uniparental disomy Autozygous segments and uniparental disomy (UPD) are diploid/'copy neutral' genetic findings and therefore are only detectable by SNP-based arrays. Both autozygous segments and UPD will show loss of heterozygosity (LOH) with a copy number of two by SNP array karyotyping. The term Runs of Homozgygosity (ROH), is a generic term that can be used for either autozygous segments or UPD. Autozygous segment: An autozygous segment is bi-parental and seen only in the germline. They are extended runs of homozygous markers in the genome, and they occur when an identical haplotype block is inherited from both parents. They are also called "identical by descent" (IBD) segments, and they can be used for homozygosity mapping.

Uniparental Disomy: UPD occurs when both copies of a gene or genomic region are inherited from the same parent. This is uniparental, in contrast to autozygous segments which are bi-parental. When present in the germline, they can be harmless or associated with disease, such as Prader-Willi or

Angelman syndromes. Also in contrast to autozygosity, UPD can develop in tumor cells, and this is referred to as acquired UPD or copy neutral LOH in the literature.

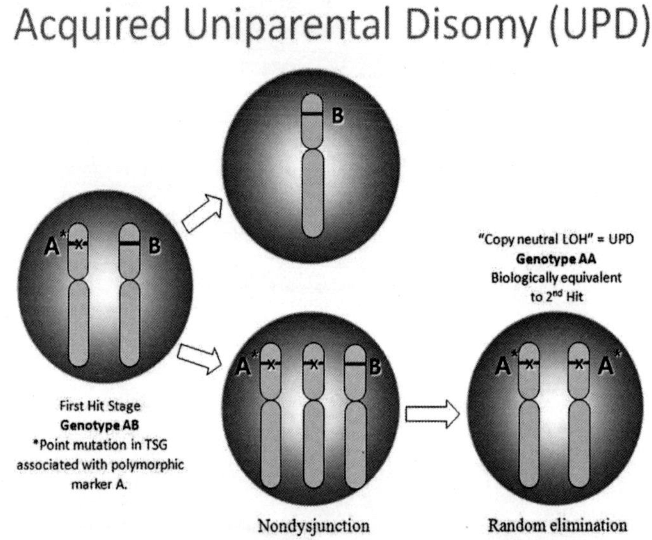

Figure: *Copy neutral LOH/uniparental disomy*

Acquired UPD is quite common in both hematologic and solid tumors, and is reported to constitute 20 to 80% of the LOH seen in human tumors. Acquired UPD can serve as the 2nd hit in the Knudson Two Hit Hypothesis of Tumorigenesis, and thus can be the biological equivalent of a deletion. Because this type of lesion cannot be detected by arrayCGH, FISH, or conventional cytogenetics, SNP-based arrays are preferred for virtual karyotyping of tumors.

Virtual karyotype of a colorectal carcinoma (whole genome view) demonstrating deletions, gains, amplifications, and acquired UPD (copy neutral LOH). is a SNP array virtual karyotype from a colorectal carcinoma demonstrating deletions, gains, amplifications, and acquired UPD (copy neutral LOH).

Examples of Clinical Cancer Applications

A virtual karyotype can be generated from nearly any tumor, but the clinical meaning of the genomic aberrations identified are different for each tumor type. Clinical utility varies and appropriateness is best determined by an oncologist or pathologist in consultation with the laboratory director of the lab performing the virtual karyotype. Below are examples of types of cancers where the clinical implications of specific genomic aberrations are well established. This list is representative, not exhaustive. The web site for the Cytogenetics Laboratory at Wisconsin State Laboratory of Hygiene has additional examples of clinically relevant genetic changes that are readily detectable by virtual karyotyping.

Neuroblastoma

Based on a series of 493 neuroblastoma samples, it has been reported that overall genomic pattern, as tested by array-based karyotyping, is a predictor of outcome in neuroblastoma:

- Tumors presenting exclusively with whole chromosome copy number changes were associated with excellent survival.
- Tumors presenting with any kind of segmental chromosome copy number changes were associated with a high risk of relapse.
- Within tumors showing segmental alterations, additional independent predictors of decreased overall survival were MYCN amplification, 1p and 11q deletions, and 1q gain.

Earlier publications categorized neuroblastomas into three major subtypes based on cytogenetic profiles:

- Subtype 1: favorable neuroblastoma with near triploidy and a predominance of numerical gains and losses, mostly representing non-metastatic NB stages 1, 2 and 4S.
- Subtypes 2A and 2B: found in unfavorable widespread neuroblastoma, stages 3 and 4, with 11q loss and 17q gain without MYCN amplification (subtype 2A) or with MYCN amplification often together with 1p deletions and 17q gain (subtype 2B).

Wilms' Tumor

Tumor-specific loss-of-heterozygosity (LOH) for chromosomes 1p and 16q identifies a subset of Wilms' tumor patients who have a significantly increased risk of relapse and death. LOH for these chromosomal regions can now be used as an independent prognostic factor together with disease stage to target intensity of treatment to risk of treatment failure.

Renal-cell Carcinoma

Renal epithelial neoplasms have characteristic cytogenetic aberrations that can aid in classification. See also Atlas of Genetics and Cytogenetics in Oncology and Haematology.

- Clear cell carcinoma: loss of 3p
- Papillary carcinoma: trisomy 7 and 17
- Chromophobe carcinoma: hypodiploid with loss of chromosomes 1, 2, 6, 10, 13, 17, 21

Array-based karyotyping can be used to identify characteristic chromosomal aberrations in renal tumors with challenging morphology. Array-based karyotyping performs well on paraffin embedded tumors and is amenable to routine clinical use.

In addition, recent literature indicates that certain chromosomal aberrations are associated with outcome in specific subtypes of renal epithelial tumors. Clear cell renal carcinoma: del 9p and del 14q are poor prognostic indicators. Papillary renal cell carcinoma: duplication of 1q marks fatal progression.

Chronic Lymphocytic Leukemia

Array-based karyotyping is a cost-effective alternative to FISH for detecting chromosomal abnormalities in chronic lymphocytic leukemia (CLL). Several clinical validation studies have shown >95% concordance with the standard CLL FISH panel. In addition, many studies using array-based karyotyping have identified 'atypical deletions' missed by the standard FISH probes and acquired uniparental disomy at key loci for prognostic risk in CLL.,

Four main genetic aberrations are recognized in CLL cells that have a major impact on disease behavior.

1. Deletions of part of the short arm of chromosome 17 (del 17p) which target p53 are particularly deleterious. Patients with this abnormality have significantly short interval before they require therapy and a shorter survival. This abnormality is found in 5–10% of patients with CLL.
2. Deletions of the long arm on chromosome 11 (del 11q) are also unfavorable although not to the degree seen with del 17p. The abnormality targets the ATM gene and occurs infrequently in CLL (5–10%).
3. Trisomy 12, an additional chromosome 12, is a relatively frequent finding occurring in 20–25% of patients and imparts an intermediate prognosis.
4. Deletion of 13q14 (del 13q14) is the most common abnormality in CLL with roughly 50% of patients with cells containing this defect. When del 13q14 is seen in isolation, patients have the best prognosis and most will live many years, even decades, without the need for therapy.

Multiple Myeloma

Avet-Loiseau, et al. in the Journal of Clinical Oncology, used SNP array karyotyping of 192 multiple myeloma (MM) samples to identify genetic lesions associated with prognosis, which were then validated in a separate cohort (n = 273). In MM, lack of a proliferative clone makes conventional cytogenetics informative in only ~30% of cases. FISH panels are useful in MM, but standard panels would not detect several key genetic abnormalities reported in this study.

1. Virtual karyotyping identified chromosomal abnormalities in 98% of MM cases
2. del(12p13.31)is an independent adverse marker
3. amp(5q31.1) is a favorable marker
4. The prognostic impact of amp(5q31.1) over-rides that of hyperdiploidy and also identifies patients who greatly benefit from high-dose therapy.

Array-based karytyping cannot detect balanced translocations, such as t(4;14) seen in ~15% of MM. Therefore, FISH for this translocation should also be performed if using SNP arrays to detect genome-wide copy number alterations of prognostic significance in MM.

Medulloblastoma

Array-based karyotyping of 260 medulloblastomas by Pfister S, et al. resulted in the following clinical subgroups based on cytogenetic profiles:

- Poor prognosis: gain of 6q or amplification of MYC or MYCN
- Intermediate: gain of 17q or an i(17q) without gain of 6q or amplfication of MYC or MYCN
- Excellent prognosis: 6q and 17q balanced or 6q deletion

Oligodendroglioma

The 1p/19q co-deletion is considered a "genetic signature" of oligodendroglioma. Allelic losses on 1p and 19q, either separately or combined, are more common in classic oligodendrogliomas than in either astrocytomas or oligoastrocytomas. In one study, classic oligodendrogliomas showed 1p loss in 35 of 42 (83%) cases, 19q loss in 28 of 39 (72%), and these were combined in 27 of 39 (69%) cases; there was no significant difference in 1p/19q loss of heterozygosity status between low-grade and anaplastic oligodendrogliomas.

1p/19q co-deletion has been correlated with both chemosensitivity and improved prognosis in oligodendrogliomas. Most larger cancer treatment centers routinely check for the deletion of 1p/19q as part of the pathology report for oligodendrogliomas. The status of the 1p/19q loci can be detected by FISH or virtual karyotyping. Virtual karyotyping has the advantage of assessing the entire genome in one assay, as well as the 1p/19q loci. This allows assessment of other key loci in glial tumors, such as EGFR and TP53 copy number status.

Whereas the prognostic relevance of 1p and 19q deletions is well established for anaplastic oligodendrogliomas and mixed oligoastrocytomas, the prognostic relevance of the deletions for low-grade gliomas is more controversial. In terms of low-grade gliomas, a recent study also suggests that 1p/19q co-deletion may be associated with a (1;19)(q10;p10) translocation which, like the combined 1p/19q deletion, is associated with superior overall survival and progression-free survival in low-grade glioma patients.

Oligodendrogliomas show only rarely mutations in the p53 gene, which is in contrast to other gliomas. Epidermal growth factor receptor amplification and whole 1p/19q codeletion are mutually exclusive and predictive of completely different outcomes, with EGFR amplification predicting poor prognosis.

Glioblastoma

Yin et al studied 55 glioblastoma and 6 GBM cell lines using SNP array karyotyping. Acquired UPD was identified at 17p in 13/61 cases. A significantly shortened survival time was found in patients with 13q14 (RB) deletion or 17p13.1 (p53) deletion/acquired UPD. Taken together,these results suggest that this technique is a rapid,robust,and inexpensive method to profile genome-wide abnormalities in GBM.Because SNP array karyotyping can be performed on paraffin embedded tumors, it is an attractive option when tumor cells fail to grow in culture for metaphase cytogenetics or when the desire for karyotyping arises after the specimen has been formalin fixed.

The importance of detecting acquired UPD (copy neutral LOH) in glioblastoma:

- Of patients with 17p abnormality, ~50% were deletions and ~50% were aUPD
- Both 17p del and 17p UPD were associated with worse outcome.
- 9/13 had homozygous TP53 mutations underlying the 17p UPD.

In addition, in cases with uncertain grade by morphology, genomic profiling can assist in diagnosis.

- Concomitant gain of 7 and loss of 10 is essentially pathognomonic for GBM
- EGFR amplification, loss of PTEN (on 10q), and loss of p16 (on 9p) occur almost exclusively in glioblastoma and can provide means to distinguish anaplastic astrocytoma from glioblastoma.

Acute Lymphoblastic Leukemia

Cytogenetics, the study of characteristic large changes in the chromosomes of cancer cells, has been increasingly recognized as an important predictor of outcome in acute lymphoblastic leukemia (ALL).

NB: Balanced translocations cannot be detected by array-based karyotyping.

Some cytogenetic subtypes have a worse prognosis than others. These include:

- A translocation between chromosomes 9 and 22, known as the Philadelphia chromosome, occurs in about 20% of adult and 5% in pediatric cases of ALL.
- A translocation between chromosomes 4 and 11 occurs in about 4% of cases and is most common in infants under 12 months.
- Not all translocations of chromosomes carry a poorer prognosis. Some translocations are relatively favorable. For example, Hyperdiploidy (>50 chromosomes) is a good prognostic factor.

- Genome-wide assessment of copy number changes can be done by conventional cytogenetics or virtual karyotyping. SNP array virtual karyotyping can detect copy number changes and LOH status, while arrayCGH can detect only copy number changes. Copy neutral LOH (acquired uniparental disomy) has been reported at key loci in ALL, such as CDKN2A gene at 9p, which have prognostic significance. SNP array virtual karyotyping can readily detect copy neutral LOH. Array CGH, FISH, and conventional cytogenetics cannot detect copy neutral LOH.

Table 1

Cytogenetic change	*Risk category*
Philadelphia chromosome	Poor prognosis
t(4;11)(q21;q23)	Poor prognosis
t(8;14)(q24.1;q32)	Poor prognosis
Complex karyotype (more than four abnormalities)	Poor prognosis
Low hypodiploidy or near triploidy	Poor prognosis
High hyperdiploidy	Good prognosis
del(9p)	Good prognosis

- Correlation of prognosis with bone marrow cytogenetic finding in acute lymphoblastic leukemia

Table 2

Prognosis	*Cytogenetic findings*
Favorable	Hyperdiploidy > 50 ; t (12;21)
Intermediate	Hyperdioloidy 47 -50; Normal(diploidy); del (6q); Rearrangements of 8q24
Unfavorable	Hypodiploidy-near haploidy; Near tetraploidy; del (17p); t (9;22); t (11q23)

- Unclassified ALL is considered to have an intermediate prognosis.

Myelodysplastic Syndrome

Myelodysplastic syndrome (MDS) has remarkable clinical, morphological, and genetic heterogeneity. Cytogenetics play a decisive role in the World Health Organization's classification-based International Prognostic Scoring System (IPSS) for MDS.

- Good Prognosis: normal karyotype, isolated del(5q), isolated del(20q), -Y
- Poor Prognosis: complex abnormalities (ie, >=3 abnormalities), "7 or del(7q)
- Intermediate Prognosis: all other abnormalities, including trisomy 8 and del(11q)

In a comparison of metaphase cytogenetics, FISH panel, and SNP array karyotyping for MDS, it was found that each technique provided a similar diagnostic yield. No single method detected all defects, and detection rates improved by ~5% when all three methods were used. Acquired UPD, which is not detectable by FISH or cytogenetics, has been reported at several key loci in MDS using SNP array karyotyping, including deletion of 7/7q.

Myeloproliferative Neoplasms/Myeloproliferative Disorders

Philadelphia chromosome–negative myeloproliferative neoplasms (MPNs) including polycythemia vera,essential thrombocythemia, and primary myelofibrosis show an inherent tendency for transformation into leukemia (MPN-blast phase),which is accompanied by acquisition of additional genomic lesions. In a study of 159 cases, SNP-array analysis was able to capture practically all cytogenetic abnormalities and to uncover additional lesions with potentially important clinical implications.

- The number of genomic alterations was more than 2 to 3 times greater in the blast phase as in the chronic phase of the disease.
- Deletion of 17p (TP53)was significantly associated with prior exposure to hydroxyurea as well as a complex karyotype in samples with MPN-blast crisis. Interestingly, not only deletion, but also 17p copy neutral LOH, was associated with a complex karyotype, a poor prognostic marker in myeloid malignancies. Copy neutral LOH (acquired UPD)is readily detectably by SNP array karyotype, but not by cytogenetics, FISH, or arrayCGH.
- Blast phase patients with loss of chromosomal material on 7q showed poor survival. Loss of 7q is known to be predictive for rapid progression and poor response in AML therapy.MPN-blast phase patients with cytogenetically undetectable 7q copy neutral-LOH had comparable survival rates to those with 7/7q in their leukemic cells.
- 9p copy neutral-LOH with homozygous JAK2 mutation was also linked to an inferior outcome in MPN-blast crisis in comparison with patients with either heterozygous JAK2V617F or wild-type JAK2. In contrast to LOH on 17p, the prognostic impact of 9pCNN-LOH was independent of established risk factors such as 7/7q, 5q, or complex karyotype.

Colorectal Cancer

Identification of biomarkers in colorectal cancer is particularly important for patients with stage II disease, where less than 20% have tumor recurrence. 18q LOH is an established biomarker associated with high risk of tumor recurrence in stage II colon cancer. Figure shows a SNP array karyotype of a colorectal carcinoma (whole genome view).

Colorectal cancers are classified into specific tumor phenotypes based on molecular profiles which can be integrated with the results of other ancillary tests, such as microsatellite instability testing, IHC, and KRAS mutation status:

- Chromosomal instability (CIN) which have allelic imbalance at a number of chromosomal loci, including 5q, 8p, 17p, and 18q.
- Microsatellite instability (MSI) which tend to have diploid karyotypes.

Malignant Rhabdoid Tumors

Malignant rhabdoid tumors are rare, highly aggressive neoplasms found most commonly in infants and young children. Due to their heterogenous histologic features, diagnosis can often be difficult and misclassifications can occur. In these tumors, the INI1 gene (SMARCB1)on chromosome 22q functions as a classic tumor suppressor gene.

Inactivation of INI1 can occur via deletion, mutation, or acquired UPD. In a recent study, SNP array karyotyping identified deletions or LOH of 22q in 49/51 rhabdoid tumors. Of these, 14 were copy neutral LOH (or acquired UPD), which is detectable by SNP array karyotyping, but not by FISH, cytogenetics, or arrayCGH. MLPA detected a single exon homozygous deletion in one sample that was below the resolution of the SNP array.

SNP array karyotyping can be used to distinguish, for example, a medulloblastoma with an isochromosome 17q from a primary rhabdoid tumor with loss of 22q11.2. When indicated, molecular analysis of INI1 using MLPA and direct sequencing may then be employed.

Once the tumor-associated changes are found, an analysis of germline DNA from the patient and the parents can be done to rule out an inherited or de novo germline mutation or deletion of INI1, so that appropriate recurrence risk assessments can be made.

Uveal Melanoma

The most important genetic alteration associated with poor prognosis in uveal melanoma is loss of an entire copy of Chromosome 3 (Monosomy 3), which is strongly correlated with metastatic spread. Gains on chromosomes 6 and 8 are often used to refine the predictive value of the Monosomy 3 screen, with gain of 6p indicating a better prognosis and gain of 8q indicating a worse prognosis in disomy 3 tumors.

In rare instances, monosomy 3 tumors may duplicate the remaining copy of the chromosome to return to a disomic state referred to as isodisomy. Isodisomy 3 is prognostically equivalent to monosomy 3, and both can be detected by tests for chromosome 3 loss of heterozygosity.

Limitations

Unlike karyotypes obtained from conventional cytogenetics, virtual karyotypes are reconstructed by computer programs using signals obtained from disrupted DNA. In essence, the computer program will correct translocations when it lines up the signals in chromosomal order. Therefore, virtual karyotypes cannot detect balanced translocations and inversions. They also can only detect genetic aberrations in regions of the genome that are represented by probes on the array. In addition, virtual karyotypes generate a relative copy number normalized against a diploid genome, so tetraploid genomes will be condensed into a diploid space unless renormalization is performed.

Renormalization requires an ancillary cell-based assay, such as FISH, if one is using arrayCGH. For karyotypes obtained from SNP-based arrays, tetraploidy can often be inferred from the maintenance of heterozygosity within a region of apparent copy number loss. Low-level mosaicism or small subclones may not be detected by virtual karyotypes because the presence of normal cells in the sample will dampen the signal from the abnormal clone.

The exact point of failure, in terms of the minimal percentage of neoplastic cells, will depend on the particular platform and algorithms used. Many copy number analysis software programs used to generate array-based karyotypes will falter with less than 25–30% tumor/abnormal cells in the sample. However, in oncology applications this limitation can be minimized by tumor enrichment strategies and/or software optimized for use with oncology samples. The analysis algorithms are evolving rapidly, and some are even designed to thrive on 'normal clone contamination', so it is anticipated that this limitation will continue to dissipate.

Happy Mapping

HAPPY Mapping, first proposed by Paul H. Dear and Peter R. Cook in 1989, is a method used in molecular biology to study the linkage between two or more DNA sequences. According to the Single Molecule Genomics Group, it is "Mapping based on the analysis of approximately HAPloid DNA samples using the PolYmerase chain reaction". In genomics, HAPPY mapping can be applied to assess the synteny and orientation of various DNA sequences across a particular genome - the generation of a "genomic" map.

As with linkage mapping, HAPPY mapping relies on the differential probability of two or more DNA sequences being separated. In genetic mapping, the probability of a recombination event between two genetic loci on the same chromosome is directly proportional to the distance between them. HAPPY mapping replaces recombination with fragmentation - instead of relying on recombination to separate genetic loci, the entire genome is fragmented, for example, by radiation or mechanical shearing. If the DNA is broken on a random basis, the longer the distance between two DNA sequences, the higher the chances of it to break between the two, and vice versa.

HAPPY mapping retains the benefits of genetic mapping while removing some of the problems associated with recombination. I.e., the need for polymorphism, and breeding. Also, recombination can be locale specific whereas breakage of genomic DNA by radiation or mechanical shearing seems to be more random.

Reverse Transcription PCR

Reverse transcription polymerase chain reaction (RT-PCR) is a variant of polymerase chain reaction (PCR), a laboratory technique commonly used in molecular biology to generate many copies of a DNA sequence, a process termed "amplification". In RT-PCR, however, an RNA strand is first reverse transcribed into its DNA complement (complementary DNA, or cDNA) using the enzyme reverse transcriptase, and the resulting cDNA is amplified using traditional PCR or real-time PCR. Reverse transcription PCR is not to be confused with real-time polymerase chain reaction (Q-PCR/qRT-PCR), which is also sometimes abbreviated as RT-PCR.

RT-PCR Principles and Procedure

RT-PCR utilizes a pair of primers, which are complementary to a defined sequence on each of the two strands of the cDNA. These primers are then extended by a DNA polymerase and a copy of the strand is made after each cycle, leading to exponential amplification.

RT-PCR includes three major steps. The first step is reverse transcription (RT), in which RNA is reverse transcribed to cDNA using reverse transcriptase. This step is very important in order to perform PCR since DNA polymerase can act only on DNA templates. The RT step can be performed either in the same tube with PCR (one-step PCR) or in a separate one (two-step PCR) using a temperature between 40°C and 50°C, depending on the properties of the reverse transcriptase used.

The next step involves the denaturation of the dsDNA at 95°C, so that the two strands separate and the primers can bind again at lower temperatures and begin a new chain reaction. Then, the temperature is decreased until it reaches the annealing temperature which can vary depending on the set of primers used, their concentration, the probe and its concentration (if used), and the cations concentration. The main consideration when choosing the optimal annealing temperature is the melting temperature (Tm) of the primers and probes (if used). The annealing temperature chosen for a PCR depends directly on length and composition of the primers. This is the result of the difference of hydrogen bonds between A-T (2 bonds) and G-C (3 bonds). An annealing temperature about 5 degrees below the lowest Tm of the pair of primers is usually used. The final step of PCR amplification is DNA extension from the primers. This is done with thermostable Taq DNA polymerase, usually at 72°C, the temperature at which the enzyme works optimally. The length of the incubation at each temperature, the temperature alterations, and the number of cycles are controlled by a programmable thermal cycler. The analysis of the PCR products depends on the type of PCR applied. If a conventional PCR is used, the PCR product is detected using agarose gel electrophoresis and ethidium bromide (or other nucleic acid staining).

Conventional RT-PCR is a time-consuming technique with important limitations when compared to real-time PCR techniques. This, combined with the fact that ethidium bromide has low sensitivity, yields results that are not always reliable. Moreover, there is an increased cross-contamination risk of the samples since detection of the PCR product requires the post-amplification processing of the samples. Furthermore, the specificity of the assay is mainly determined by the primers, which can give false-positive results. However, the most important issue concerning conventional RT-PCR is the fact that it is a semi- or even a low-quantitative technique, whereas the amplicon can be visualized only after the amplification ends.

Real-time RT-PCR provides a method in which the amplicons can be visualized as the amplification progresses using a fluorescent reporter molecule. There are three major kinds of fluorescent reporters used in real time RT-PCR, which are general non-specific DNA Binding Dyes such as SYBR Green I, TaqMan Probes and Molecular Beacons (including Scorpions). The real-time PCR thermal cycler has a fluorescence detection threshold, below which it cannot discriminate the difference between an amplification generated signal and background noise. On the other hand, the fluorescence increases as the amplification progresses and the instrument performs data acquisition during the annealing step of each cycle. The number of amplicons will reach the detection baseline after a specific cycle, which depends on the initial concentration of the target DNA sequence. The cycle at which the instrument can discriminate the amplification generated fluorescence from the background noise is called the threshold cycle (Ct). The higher the initial DNA concentration, the lower its Ct will be.

Use of reverse transcription polymerase chain reaction

The exponential amplification via reverse transcription polymerase chain reaction provides for a highly sensitive technique in which a very low copy number of RNA molecules can be detected. RT-PCR is widely used in the diagnosis of genetic diseases and, semiquantitatively, in the determination of the abundance of specific different RNA molecules within a cell or tissue as a measure of gene expression. Northern blot analysis is used to study the RNA's gene expression further. RT-PCR can also be very useful in the insertion of eukaryotic genes into prokaryotes. Because most eukaryotic genes contain introns which are present in the genome but not in the mature mRNA, the cDNA generated from a RT-PCR reaction is the exact (without regard to the error prone nature of reverse transcriptases) DNA sequence which would be directly translated into protein after transcription. When these genes are expressed in prokaryotic cells for the sake of protein production or purification, the RNA produced directly from transcription need not undergo splicing as the transcript contains only exons (prokaryotes, such as E.coli, lack the mRNA splicing mechanism of eukaryote. RT-PCR is commonly used in studying the genomes of viruses whose genomes are composed of RNA, such as Influenzavirus A and retroviruses like HIV.

Polymerase Chain Reaction

The polymerase chain reaction (PCR) is a scientific technique in molecular biology to amplify a single or a few copies of a piece of DNA across several orders of magnitude, generating thousands to millions of copies of a particular DNA sequence. Developed in 1983 by Kary Mullis, PCR is now a common and often indispensable technique used in medical and biological research labs for a variety of applications. These include DNA cloning for sequencing, DNA-based phylogeny, or functional analysis of genes; the diagnosis of hereditary diseases; the identification of genetic fingerprints (used in forensic sciences and paternity testing); and the detection and diagnosis of infectious diseases. In 1993, Mullis was awarded the Nobel Prize in Chemistry along with Michael Smith for his work on PCR.

The method relies on thermal cycling, consisting of cycles of repeated heating and cooling of the reaction for DNA melting and enzymatic replication of the DNA. Primers (short DNA fragments) containing sequences complementary to the target region along with a DNA polymerase (after which the method is named) are key components to enable selective and repeated amplification. As PCR progresses, the DNA generated is itself used as a template for replication, setting in motion a chain reaction in which the DNA template is exponentially amplified. PCR can be extensively modified to perform a wide array of genetic manipulations. Almost all PCR applications employ a heat-stable DNA polymerase, such as Taq polymerase, an enzyme originally isolated from the bacterium Thermus aquaticus. This DNA polymerase enzymatically assembles a new DNA strand from DNA building-blocks, the nucleotides, by using single-stranded DNA as a template and DNA oligonucleotides (also called DNA primers), which are required for initiation of DNA synthesis. The vast majority of PCR methods use thermal cycling, i.e., alternately heating and cooling the PCR sample to a defined series of temperature steps. These thermal cycling steps are necessary first to physically separate the two strands in a DNA double helix at a high temperature in a process called DNA melting. At a lower temperature, each strand is then used as the template in DNA synthesis by the DNA polymerase to selectively amplify the target DNA. The selectivity of PCR results from the use of primers that are complementary to the DNA region targeted for amplification under specific thermal cycling conditions.

PCR Principles and Procedure

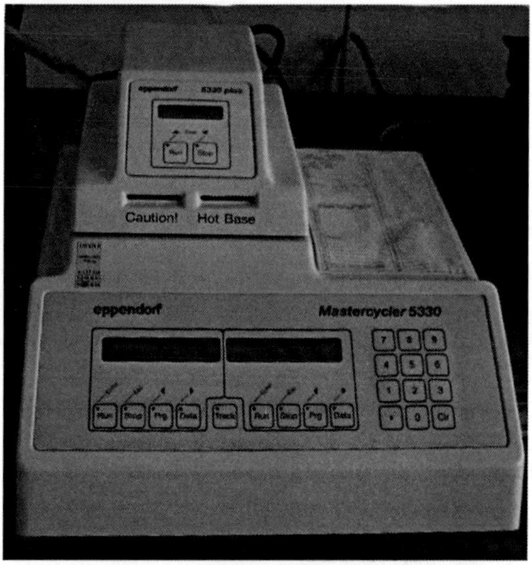

Figure: *A thermal cycler for PCR*

Figure: *An older model three-temperature thermal cycler for PCR*

PCR is used to amplify a specific region of a DNA strand (the DNA target). Most PCR methods typically amplify DNA fragments of up to ~10 kilo base pairs (kb), although some techniques allow for amplification of fragments up to 40 kb in size.

A basic PCR set up requires several components and reagents. These components include:

- DNA template that contains the DNA region (target) to be amplified.
- Two primers that are complementary to the 3' (three prime) ends of each of the sense and anti-sense strand of the DNA target.
- Taq polymerase or another DNA polymerase with a temperature optimum at around 70 °C.
- Deoxynucleoside triphosphates (dNTPs; nucleotides containing triphosphate groups), the building-blocks from which the DNA polymerase synthesizes a new DNA strand.
- Buffer solution, providing a suitable chemical environment for optimum activity and stability of the DNA polymerase.
- Divalent cations, magnesium or manganese ions; generally Mg is used, but Mn can be utilized for PCR-mediated DNA mutagenesis, as higher Mn concentration increases the error rate during DNA synthesis
- Monovalent cation potassium ions.

The PCR is commonly carried out in a reaction volume of 10–200 ìl in small reaction tubes (0.2–0.5 ml volumes) in a thermal cycler. The thermal cycler heats and cools the reaction tubes to achieve the temperatures required at each step of the reaction . Many modern thermal cyclers make use of the Peltier effect, which permits both heating and cooling of the block holding the PCR tubes simply by reversing the electric current.

Thin-walled reaction tubes permit favorable thermal conductivity to allow for rapid thermal equilibration. Most thermal cyclers have heated lids to prevent condensation at the top of the reaction tube. Older thermocyclers lacking a heated lid require a layer of oil on top of the reaction mixture or a ball of wax inside the tube.

Procedure

Typically, PCR consists of a series of 20-40 repeated temperature changes, called cycles, with each cycle commonly consisting of 2-3 discrete temperature steps, usually three. The cycling is often preceded by a single temperature step (called hold) at a high temperature (>90°C), and followed by one hold at the end for final product extension or brief storage. The temperatures used and the length of time they are applied in each cycle depend on a variety of parameters. These include the enzyme used for DNA synthesis, the concentration of divalent ions and dNTPs in the reaction, and the melting temperature (Tm) of the primers.

- Initialization step: This step consists of heating the reaction to a temperature of 94–96 °C (or 98 °C if extremely thermostable polymerases are used), which is held for 1–9 minutes. It is only required for DNA polymerases that require heat activation by hot-start PCR.
- Denaturation step: This step is the first regular cycling event and consists of heating the reaction to 94–98 °C for 20–30 seconds. It causes DNA melting of the DNA template by disrupting the hydrogen bonds between complementary bases, yielding single-stranded DNA molecules.
- Annealing step: The reaction temperature is lowered to 50–65 °C for 20–40 seconds allowing annealing of the primers to the single-stranded DNA template. Typically the annealing temperature is about 3-5 degrees Celsius below the Tm of the primers used. Stable DNA-DNA hydrogen bonds are only formed when the primer sequence very closely matches the template sequence. The polymerase binds to the primer-template hybrid and begins DNA synthesis.
- Extension/elongation step: The temperature at this step depends on the DNA polymerase used; Taq polymerase has its optimum activity temperature at 75–80 °C, and commonly a temperature of 72 °C is used with this enzyme. At this step the DNA polymerase synthesizes a new DNA strand complementary to the DNA template strand by adding dNTPs that are complementary to the template in 5' to 3' direction, condensing the 5'-phosphate group of the dNTPs with the 3'-hydroxyl group at the end of the nascent (extending) DNA strand. The extension time depends both on the DNA polymerase used and on the length of the DNA fragment to be amplified. As a rule-of-thumb, at its optimum temperature, the DNA polymerase will polymerize a thousand bases per minute. Under optimum conditions, i.e., if there are no limitations due to limiting substrates or reagents, at each extension step, the amount of DNA target is doubled, leading to exponential (geometric) amplification of the specific DNA fragment.
- Final elongation: This single step is occasionally performed at a temperature of 70–74 °C for 5–15 minutes after the last PCR cycle to ensure that any remaining single-stranded DNA is fully extended.
- Final hold: This step at 4–15 °C for an indefinite time may be employed for short-term storage of the reaction.

PCR Stages

The PCR process can be divided into three stages:

Exponential amplification: At every cycle, the amount of product is doubled (assuming 100% reaction efficiency). The reaction is very sensitive: only minute quantities of DNA need to be present.

Leveling off stage: The reaction slows as the DNA polymerase loses activity and as consumption of reagents such as dNTPs and primers causes them to become limiting.

Plateau: No more product accumulates due to exhaustion of reagents and enzyme.

PCR Optimization

In practice, PCR can fail for various reasons, in part due to its sensitivity to contamination causing amplification of spurious DNA products. Because of this, a number of techniques and procedures have been developed for optimizing PCR conditions. Contamination with extraneous DNA is addressed with lab protocols and procedures that separate pre-PCR mixtures from potential DNA contaminants. This usually involves spatial separation of PCR-setup areas from areas for analysis or purification of PCR products, use of disposable plasticware, and thoroughly cleaning the work surface between reaction setups.

Primer-design techniques are important in improving PCR product yield and in avoiding the formation of spurious products, and the usage of alternate buffer components or polymerase enzymes can help with amplification of long or otherwise problematic regions of DNA. Addition of reagents, such as formamide, in buffer systems may increase the specificity and yield of PCR.

Application of PCR

Selective DNA Isolation

PCR allows isolation of DNA fragments from genomic DNA by selective amplification of a specific region of DNA. This use of PCR augments many methods, such as generating hybridization probes for Southern or northern hybridization and DNA cloning, which require larger amounts of DNA, representing a specific DNA region. PCR supplies these techniques with high amounts of pure DNA, enabling analysis of DNA samples even from very small amounts of starting material.

Other applications of PCR include DNA sequencing to determine unknown PCR-amplified sequences in which one of the amplification primers may be used in Sanger sequencing, isolation of a DNA sequence to expedite recombinant DNA technologies involving the insertion of a DNA sequence into a plasmid or the genetic material of another organism. Bacterial colonies (E. coli) can be rapidly screened by PCR for correct DNA vector constructs. PCR may also be used for genetic fingerprinting; a forensic technique used to identify a person or organism by comparing experimental DNAs through different PCR-based methods. Some PCR 'fingerprints' methods have high discriminative power and can be used to identify genetic relationships between individuals, such as parent-child or between siblings, and are used in paternity testing. This technique may also be used to determine evolutionary relationships among organisms.

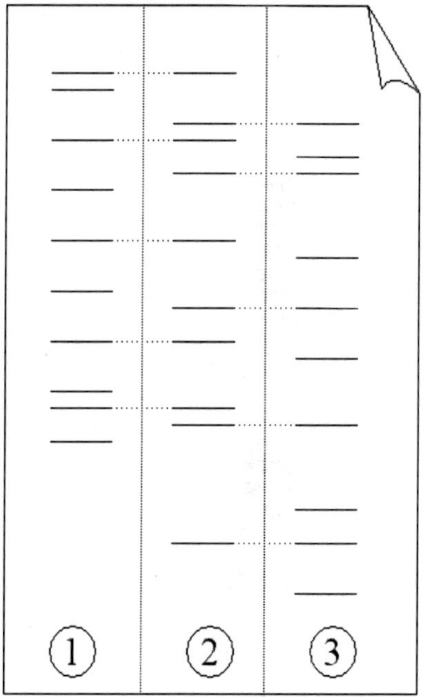

Figure: *Electrophoresis of PCR-amplified DNA fragments. (1) Father. (2) Child. (3) Mother. The child has inherited some, but not all of the fingerprint of each of its parents, giving it a new, unique fingerprint.*

Amplification and Quantification of DNA

Because PCR amplifies the regions of DNA that it targets, PCR can be used to analyze extremely small amounts of sample. This is often critical for forensic analysis, when only a trace amount of DNA is available as evidence. PCR may also be used in the analysis of ancient DNA that is tens of thousands of years old. These PCR-based techniques have been successfully used on animals, such as a forty-thousand-year-old mammoth, and also on human DNA, in applications ranging from the analysis of Egyptian mummies to the identification of a Russian tsar. Quantitative PCR methods allow the estimation of the amount of a given sequence present in a sample—a technique often applied to quantitatively determine levels of gene expression. Real-time PCR is an established tool for DNA quantification that measures the accumulation of DNA product after each round of PCR amplification.

DNA in Forensic Entomology

PCR in Diagnosis of Diseases

PCR permits early diagnosis of malignant diseases such as leukemia and lymphomas, which is currently the highest-developed in cancer research and is already being used routinely. PCR assays can be performed directly on genomic DNA samples to detect translocation-specific malignant cells at a sensitivity that is at least 10,000-fold higher than that of other methods.

PCR also permits identification of non-cultivatable or slow-growing microorganisms such as mycobacteria, anaerobic bacteria, or viruses from tissue culture assays and animal models. The basis for PCR diagnostic applications in microbiology is the detection of infectious agents and the discrimination of non-pathogenic from pathogenic strains by virtue of specific genes.

Viral DNA can likewise be detected by PCR. The primers used need to be specific to the targeted sequences in the DNA of a virus, and the PCR can be used for diagnostic analyses or DNA sequencing of the viral genome. The high sensitivity of PCR permits virus detection soon after infection and even before the onset of disease. Such early detection may give physicians a significant lead in treatment. The amount of virus ("viral load") in a patient can also be quantified by PCR-based DNA quantitation techniques .

Variations on the Basic PCR Technique

- Allele-specific PCR: a diagnostic or cloning technique based on single-nucleotide polymorphisms (SNPs) (single-base differences in DNA). It requires prior knowledge of a DNA sequence, including differences between alleles, and uses primers whose 3' ends encompass the SNP. PCR amplification under stringent conditions is much less efficient in the presence of a mismatch between template and primer, so successful amplification with an SNP-specific primer signals presence of the specific SNP in a sequence.
- Assembly PCR or Polymerase Cycling Assembly (PCA): artificial synthesis of long DNA sequences by performing PCR on a pool of long oligonucleotides with short overlapping segments. The oligonucleotides alternate between sense and antisense directions, and the overlapping segments determine the order of the PCR fragments, thereby selectively producing the final long DNA product.
- Asymmetric PCR: preferentially amplifies one DNA strand in a double-stranded DNA template. It is used in sequencing and hybridization probing where amplification of only one of the two complementary strands is required. PCR is carried out as usual, but with a great excess of the primer for the strand targeted for amplification. Because of the

slow (arithmetic) amplification later in the reaction after the limiting primer has been used up, extra cycles of PCR are required. A recent modification on this process, known as Linear-After-The-Exponential-PCR (LATE-PCR), uses a limiting primer with a higher melting temperature (Tm) than the excess primer to maintain reaction efficiency as the limiting primer concentration decreases mid-reaction.

- Helicase-dependent amplification: similar to traditional PCR, but uses a constant temperature rather than cycling through denaturation and annealing/extension cycles. DNA helicase, an enzyme that unwinds DNA, is used in place of thermal denaturation.
- Hot start PCR: a technique that reduces non-specific amplification during the initial set up stages of the PCR. It may be performed manually by heating the reaction components to the denaturation temperature (e.g., 95°C) before adding the polymerase. Specialized enzyme systems have been developed that inhibit the polymerase's activity at ambient temperature, either by the binding of an antibody or by the presence of covalently bound inhibitors that dissociate only after a high-temperature activation step. Hot-start/cold-finish PCR is achieved with new hybrid polymerases that are inactive at ambient temperature and are instantly activated at elongation temperature.
- Intersequence-specific PCR (ISSR): a PCR method for DNA fingerprinting that amplifies regions between simple sequence repeats to produce a unique fingerprint of amplified fragment lengths.
- Inverse PCR: is commonly used to identify the flanking sequences around genomic inserts. It involves a series of DNA digestions and self ligation, resulting in known sequences at either end of the unknown sequence.
- Ligation-mediated PCR: uses small DNA linkers ligated to the DNA of interest and multiple primers annealing to the DNA linkers; it has been used for DNA sequencing, genome walking, and DNA footprinting.
- Methylation-specific PCR (MSP): developed by Stephen Baylin and Jim Herman at the Johns Hopkins School of Medicine, and is used to detect methylation of CpG islands in genomic DNA. DNA is first treated with sodium bisulfite, which converts unmethylated cytosine bases to uracil, which is recognized by PCR primers as thymine. Two PCRs are then carried out on the modified DNA, using primer sets identical except at any CpG islands within the primer sequences. At these points, one primer set recognizes DNA with cytosines to amplify methylated DNA, and one set recognizes DNA with uracil or thymine to amplify

unmethylated DNA. MSP using qPCR can also be performed to obtain quantitative rather than qualitative information about methylation.

- Miniprimer PCR: uses a thermostable polymerase (S-Tbr) that can extend from short primers ("smalligos") as short as 9 or 10 nucleotides. This method permits PCR targeting to smaller primer binding regions, and is used to amplify conserved DNA sequences, such as the 16S (or eukaryotic 18S) rRNA gene.
- Multiplex Ligation-dependent Probe Amplification (MLPA): permits multiple targets to be amplified with only a single primer pair, thus avoiding the resolution limitations of multiplex PCR .
- Multiplex-PCR: consists of multiple primer sets within a single PCR mixture to produce amplicons of varying sizes that are specific to different DNA sequences. By targeting multiple genes at once, additional information may be gained from a single test-run that otherwise would require several times the reagents and more time to perform. Annealing temperatures for each of the primer sets must be optimized to work correctly within a single reaction, and amplicon sizes. That is, their base pair length should be different enough to form distinct bands when visualized by gel electrophoresis.
- Nested PCR: increases the specificity of DNA amplification, by reducing background due to non-specific amplification of DNA. Two sets of primers are used in two successive PCRs. In the first reaction, one pair of primers is used to generate DNA products, which besides the intended target, may still consist of non-specifically amplified DNA fragments. The product(s) are then used in a second PCR with a set of primers whose binding sites are completely or partially different from and located 3' of each of the primers used in the first reaction. Nested PCR is often more successful in specifically amplifying long DNA fragments than conventional PCR, but it requires more detailed knowledge of the target sequences.
- Overlap-extension PCR or Splicing by overlap extension (SOE) : a genetic engineering technique that is used to splice together two or more DNA fragments that contain complementary sequences. It is used to join DNA pieces containing genes, regulatory sequences, or mutations; the technique enables creation of specific and long DNA constructs.
- Quantitative PCR (Q-PCR): used to measure the quantity of a PCR product (commonly in real-time). It quantitatively measures starting amounts of DNA, cDNA, or RNA. Q-PCR is commonly used to determine whether a DNA sequence is present in a sample and the number of

its copies in the sample. Quantitative real-time PCR has a very high degree of precision. QRT-PCR (or QF-PCR) methods use fluorescent dyes, such as Sybr Green, EvaGreen or fluorophore-containing DNA probes, such as TaqMan, to measure the amount of amplified product in real time. It is also sometimes abbreviated to RT-PCR (Real Time PCR) or RQ-PCR. QRT-PCR or RTQ-PCR are more appropriate contractions, since RT-PCR commonly refers to reverse transcription PCR, often used in conjunction with Q-PCR.

- Reverse Transcription PCR (RT-PCR): for amplifying DNA from RNA. Reverse transcriptase reverse transcribes RNA into cDNA, which is then amplified by PCR. RT-PCR is widely used in expression profiling, to determine the expression of a gene or to identify the sequence of an RNA transcript, including transcription start and termination sites. If the genomic DNA sequence of a gene is known, RT-PCR can be used to map the location of exons and introns in the gene. The 5' end of a gene (corresponding to the transcription start site) is typically identified by RACE-PCR (Rapid Amplification of cDNA Ends).
- Solid Phase PCR: encompasses multiple meanings, including Polony Amplification (where PCR colonies are derived in a gel matrix, for example), Bridge PCR (primers are covalently linked to a solid-support surface), conventional Solid Phase PCR (where Asymmetric PCR is applied in the presence of solid support bearing primer with sequence matching one of the aqueous primers) and Enhanced Solid Phase PCR (where conventional Solid Phase PCR can be improved by employing high Tm and nested solid support primer with optional application of a thermal 'step' to favour solid support priming).
- Thermal asymmetric interlaced PCR (TAIL-PCR): for isolation of an unknown sequence flanking a known sequence. Within the known sequence, TAIL-PCR uses a nested pair of primers with differing annealing temperatures; a degenerate primer is used to amplify in the other direction from the unknown sequence.
- Touchdown PCR (Step-down PCR): a variant of PCR that aims to reduce nonspecific background by gradually lowering the annealing temperature as PCR cycling progresses. The annealing temperature at the initial cycles is usually a few degrees (3-5°C) above the T_m of the primers used, while at the later cycles, it is a few degrees (3-5°C) below the primer T_m. The higher temperatures give greater specificity for primer binding, and the lower temperatures permit more efficient amplification from the specific products formed during the initial cycles.

- PAN-AC: uses isothermal conditions for amplification, and may be used in living cells.
- Universal Fast Walking: for genome walking and genetic fingerprinting using a more specific 'two-sided' PCR than conventional 'one-sided' approaches (using only one gene-specific primer and one general primer — which can lead to artefactual 'noise') by virtue of a mechanism involving lariat structure formation. Streamlined derivatives of UFW are LaNe RAGE (lariat-dependent nested PCR for rapid amplification of genomic DNA ends), 5'RACE LaNe and 3'RACE LaNe.

History

A 1971 paper in the Journal of Molecular Biology by Kleppe and co-workers first described a method using an enzymatic assay to replicate a short DNA template with primers in vitro. However, this early manifestation of the basic PCR principle did not receive much attention, and the invention of the polymerase chain reaction in 1983 is generally credited to Kary Mullis.

When Mullis developed the PCR in 1983, he was working in Emeryville, California for Cetus Corporation, one of the first biotechnology companies. There, he was responsible for synthesizing short chains of DNA. Mullis has written that he conceived of PCR while cruising along the Pacific Coast Highway one night in his car. He was playing in his mind with a new way of analyzing changes (mutations) in DNA when he realized that he had instead invented a method of amplifying any DNA region through repeated cycles of duplication driven by DNA polymerase. In Scientific American, Mullis summarized the procedure:

"Beginning with a single molecule of the genetic material DNA, the PCR can generate 100 billion similar molecules in an afternoon. The reaction is easy to execute. It requires no more than a test tube, a few simple reagents, and a source of heat." He was awarded the Nobel Prize in Chemistry in 1993 for his invention, seven years after he and his colleagues at Cetus first put his proposal to practice. However, some controversies have remained about the intellectual and practical contributions of other scientists to Mullis' work, and whether he had been the sole inventor of the PCR principle.

At the core of the PCR method is the use of a suitable DNA polymerase able to withstand the high temperatures of >90 °C (194 °F) required for separation of the two DNA strands in the DNA double helix after each replication cycle. The DNA polymerases initially employed for in vitro experiments presaging PCR were unable to withstand these high temperatures. So the early procedures for DNA replication were very inefficient and time consuming, and required large amounts of DNA polymerase and continuous handling throughout the process.

The discovery in 1976 of Taq polymerase — a DNA polymerase purified from the thermophilic bacterium, Thermus aquaticus, which naturally lives in hot (50 to 80 °C (122 to 176 °F)) environments such as hot springs — paved the way for dramatic improvements of the PCR method. The DNA polymerase isolated from T. aquaticus is stable at high temperatures remaining active even after DNA denaturation, thus obviating the need to add new DNA polymerase after each cycle. This allowed an automated thermocycler-based process for DNA amplification.

Patent Wars

The PCR technique was patented by Kary Mullis and assigned to Cetus Corporation, where Mullis worked when he invented the technique in 1983. The Taq polymerase enzyme was also covered by patents. There have been several high-profile lawsuits related to the technique, including an unsuccessful lawsuit brought by DuPont. The pharmaceutical company Hoffmann-La Roche purchased the rights to the patents in 1992 and currently holds those that are still protected.

A related patent battle over the Taq polymerase enzyme is still ongoing in several jurisdictions around the world between Roche and Promega. The legal arguments have extended beyond the lives of the original PCR and Taq polymerase patents, which expired on March 28, 2005.

NanoString Technologies

NanoString Technologies is a privately held life sciences company that develops and manufactures solutions for detecting and counting large sets of target molecules in biological samples. The company was founded by Dr. Krassen Dimitrov in 2003 and is based in Seattle, WA.

The company's products are based on a digital molecular barcoding technology invented by Dr. Dimitrov at the Institute for Systems Biology (ISB) in Seattle in the lab of Dr. Leroy Hood. This technology underlies the company's nCounter™ Analysis System which became commercially available worldwide in 2008. NanoString's products are used for a variety of discovery and translational research applications including gene expression, miRNA analysis, and copy number variation analysis. NanoString is also developing the technology for use in molecular diagnostics. The company's investors include Clarus Ventures, OVP Venture Partners and Draper Fisher Jurvetson.

The Technology

The Nanostring technology was invented by Dr. Krassen Dimitrov, and is protected by U.S. Patent 7,473,767. nCounter Analysis System utilizes a digital technology that is based on direct multiplexed measurement of gene expression that is capable of high level precision and sensitivity at less than

1 transcript copy per cell. The technology uses molecular "barcodes" and single-molecule imaging to detect and count hundreds of unique transcripts in a single reaction.

Each color-coded barcode is attached to a single target-specific probe corresponding to a gene of interest. Mixed together with controls, they form a multiplexed assay. The degree of multiplexing is in the hundreds, which is less than that of microarrays. However, the nanostring technology has higher throughput, accuracy and sensitivity than microarrays, which makes it preferable for low-multiplex applications, such as biomarker validation or molecular diagnostics.

The nanostring process includes the following steps:

- Hybridization: NanoString's Technology employs two ~50 base probes per mRNA that hybridize in solution. The Reporter Probe carries the signal; the Capture Probe allows the complex to be immobilized for data collection.
- Purification and Immobilization: After hybridization, the excess probes are removed and the probe/target complexes are aligned and immobilized in the nCounter Cartridge.
- Data Collection: Sample Cartridges are placed in the Digital Analyzer instrument for data collection. Color codes on the surface of the cartridge are counted and tabulated for each target molecule.

The nCounter Analysis System primarily addresses the target validation and routine testing segments of the genomics research market in both academia and the pharmaceutical industry. Cancer research and biomarker validation are two of the areas where NanoString has seen the most rapid adoption of its nCounter Analysis System.

Products

NanoString's products include:

The nCounter Analysis System: The system consists of two instruments; the Prep Station which is an automated fluidic instrument that immobilizes CodeSet complexes for data collection and the Digital Analyzer which derives data by counting of fluorescent barcodes.

CodeSets: CodeSets are custom-made or pre-designed sets of color-coded probes pre-mixed with a set of system controls for use in a variety of basic research and translational medicine applications including gene expression, miRNA analysis, and copy number variation.

- Higher-plex techniques:
  - o SAGE

Serial analysis of gene expression (SAGE) is a technique used by molecular biologists to produce a snapshot of the messenger RNA population in a sample of interest in the form of small tags that correspond to fragments of those transcripts. The original technique was developed by Dr. Victor Velculescu at the Oncology Center of Johns Hopkins University and published in 1995. Several variants have been developed since, most notably a more robust version, LongSAGE, RL-SAGE and the most recent SuperSAGE. Many of these have improved the technique with the capture of longer tags, enabling more confident identification of a source gene.

Overview

Briefly, SAGE experiments proceed as follows:

1. Isolate the mRNA of an input sample (e.g. a tumour).
2. Extract a small chunk of sequence from a defined position of each mRNA molecule.
3. Link these small pieces of sequence together to form a long chain (or concatemer).
4. Clone these chains into a vector which can be taken up by bacteria.
5. Sequence these chains using modern high-throughput DNA sequencers.
6. Process this data with a computer to count the small sequence tags.

A more in-depth, technical explanation of the technique is available here.

Analysis

The output of SAGE is a list of short sequence tags and the number of times it is observed. Using sequence databases a researcher can usually determine, with some confidence, the original mRNA (and therefore which gene) the tag was extracted from. Statistical methods can be applied to tag and count lists from different samples in order to determine which genes are more highly expressed. For example, a normal tissue sample can be compared against a corresponding tumour to determine which genes tend to be more (or less) active.

Applications

Although SAGE was originally conceived for use in cancer studies, it has been successfully used to describe the transcriptome of other diseases and in a wide variety of organisms.

Comparison to DNA Microarrays

The general goal of the technique is similar to the DNA microarray. However, SAGE sampling is based on sequencing mRNA output, not on hybridization of mRNA output to probes, so transcription levels are measured more quantitativly than by microarray. In addition, the mRNA sequences do not need to be known a priori, so genes or gene variants which are not known can be discovered. Microarray experiments are much cheaper to perform, so large-scale studies do not typically use SAGE. Quantifying gene expressions is more exact in SAGE because it involves directly counting the number of transcripts whereas spot intensities in microarrays fall in non-discrete gradients and are prone to background noise.

Variant Protocols: miRNA Cloning

MicroRNAs, or miRNAs for short, are small (~22nt) segments of RNA which have been found to play a crucial role in gene regulation. One of the most commonly used methods for cloning and identifying miRNAs within a cell or tissue was developed in the Bartel Lab and published in a paper by Lau et al. (2001). Since then, several variant protocols have arisen, but most have the same basic format. The procedure is quite similar to SAGE: The small RNA are isolated, then linkers are added to each, and the RNA is converted to cDNA by RT-PCR. Following this, the linkers, containing internal restriction sites, are digested with the appropriate restriction enzyme and the sticky ends are ligated together into concatamers. Following concatenation, the fragments are ligated into plasmids and are used to transform bacteria to generate many copies of the plasmid containing the inserts. Those may then be sequenced to identify the miRNA present, as well as analysing expression levels of a given miRNA by counting the number of times it is present, similar to SAGE.

DNA Microarray

A DNA microarray (also commonly known as gene chip, DNA chip, or biochip) is a collection of microscopic DNA spots attached to a solid surface. Scientists use DNA microarrays to measure the expression levels of large numbers of genes simultaneously or to genotype multiple regions of a genome. Each DNA spot contains picomoles (10 moles) of a specific DNA sequence, known as probes (or reporters). These can be a short section of a gene or other DNA element that are used to hybridize a cDNA or cRNA sample (called target) under high-stringency conditions. Probe-target hybridization is usually detected and quantified by detection of fluorophore-, silver-, or chemiluminescence-labeled targets to determine relative abundance of nucleic acid sequences in the target.

3

The Basic Microarray

Since an array can contain tens of thousands of probes, a microarray experiment can accomplish many genetic tests in parallel. Therefore arrays have dramatically accelerated many types of investigation. In standard microarrays, the probes are synthesized and then attached via surface engineering to a solid surface by a covalent bond to a chemical matrix (via epoxy-silane, amino-silane, lysine, polyacrylamide or others). The solid surface can be glass or a silicon chip, in which case they are colloquially known as an Affy chip when an Affymetrix chip is used. Other microarray platforms, such as Illumina, use microscopic beads, instead of the large solid support. Alternatively, microarrays can be constructed by the direct synthesis of oligonucleotide probes on solid surfaces. DNA arrays are different from other types of microarray only in that they either measure DNA or use DNA as part of its detection system. DNA microarrays can be used to measure changes in expression levels, to detect single nucleotide polymorphisms (SNPs), or to genotype or resequence mutant genomes. Microarrays also differ in fabrication, workings, accuracy, efficiency, and cost. Additional factors for microarray experiments are the experimental design and the methods of analyzing the data.

History

Microarray technology evolved from Southern blotting, where fragmented DNA is attached to a substrate and then probed with a known gene or fragment. The first reported use of this approach was the analysis of 378 arrayed lysed bacterial colonies each harboring a different sequence which were assayed in multiple replicas for expression of the genes in multiple normal and tumor tissue. This was expanded to analysis of more than 4000 human sequences with computer driven scanning and image processing for quantitative analysis of the sequences in human colonic tumors and normal tissue and then to comparison of colonic tissues at different genetic risk. The use of a collection of distinct DNAs in arrays for expression profiling was also described in 1987, and the arrayed DNAs were used to identify genes whose expression is modulated by interferon. These early gene arrays were made by spotting cDNAs onto filter paper with a pin-spotting device. The use of miniaturized microarrays for gene expression profiling was first reported in 1995, and a complete eukaryotic genome (Saccharomyces cerevisiae) on a microarray was published in 1997.

Principle

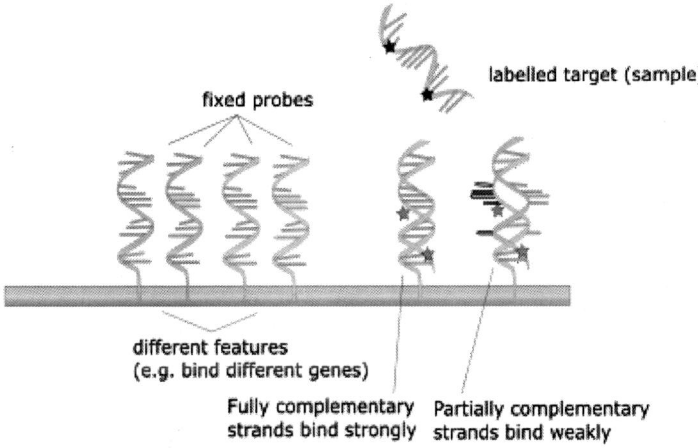

Figure: *Hybridization of the target to the probe.*

The core principle behind microarrays is hybridization between two DNA strands, the property of complementary nucleic acid sequences to specifically pair with each other by forming hydrogen bonds between complementary nucleotide base pairs. A high number of complementary base pairs in a nucleotide sequence means tighter non-covalent bonding between the two strands. After washing off of non-specific bonding sequences, only strongly paired strands will remain hybridized. So fluorescently labeled target sequences that bind to a probe sequence generate a signal that depends on the strength of the hybridization determined by the number of paired bases, the hybridization conditions (such as temperature), and washing after hybridization. Total strength of the signal, from a spot (feature), depends upon the amount of target sample binding to the probes present on that spot. Microarrays use relative quantization in which the intensity of a feature is compared to the intensity of the same feature under a different condition, and the identity of the feature is known by its position.

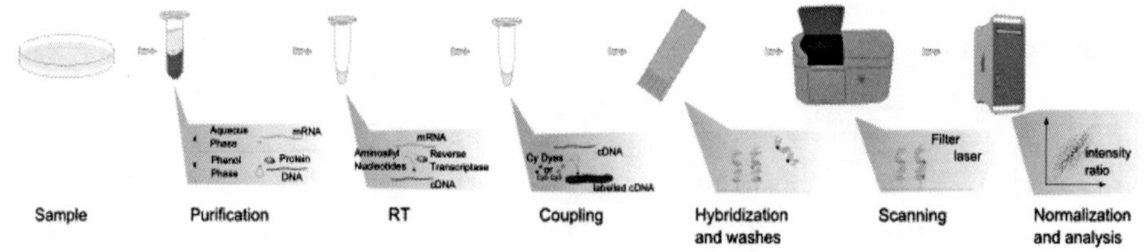

Figure: *The step required in a microarray experiment.*

Uses and Types

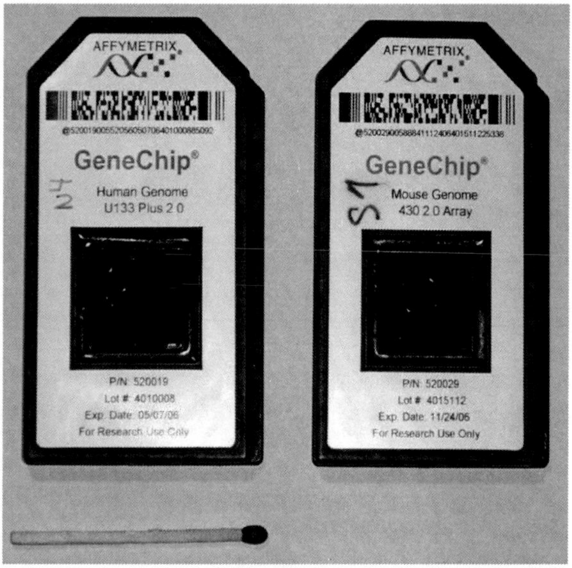

Figure: *Two Affymetrix chips. A match is shown at bottom left for size comparison.*

Many types of array exist and the broadest distinction is whether they are spatially arranged on a surface or on coded beads:

- The traditional solid-phase array is a collection of orderly microscopic "spots", called features, each with a specific probe attached to a solid surface, such as glass, plastic or silicon biochip (commonly known as a genome chip, DNA chip or gene array). Thousands of them can be placed in known locations on a single DNA microarray.
- The alternative bead array is a collection of microscopic polystyrene beads, each with a specific probe and a ratio of two or more dyes, which do not interfere with the fluorescent dyes used on the target sequence.

DNA microarrays can be used to detect DNA (as in comparative genomic hybridization), or detect RNA (most commonly as cDNA after reverse transcription) that may or may not be translated into proteins. The process of measuring gene expression via cDNA is called expression analysis or expression profiling.

***Table:** Applications include:*

Application or technology	Synopsis
Gene expression	In an mRNA or gene expression profiling experiment the profiling expression levels of thousands of genes are simultaneously monitored to study the effects of certain treatments, diseases, and developmental stages on gene expression. For example, microarray-based gene expression profiling can be used to identify genes whose expression is changed in response to pathogens or other organisms by comparing gene expression in infected to that in uninfected cells or tissues.
Comparative geno	Assessing genome content in different cells or closely related -mic hybridization organisms.
GeneID	Small microarrays to check IDs of organisms in food and feed (like GMO), mycoplasms in cell culture, or pathogens for disease detection, mostly combining PCR and microarray technology.
Chromatin immunop-	DNA sequences bound to a particular protein can be isolated recipitation on Chipby immunoprecipitating that protein (ChIP), these fragments can be then hybridized to a microarray (such as a tiling array) allowing the determination of protein binding site occupancy throughout the genome. Example protein to immunoprecipitate are histone modifications (H3K27me3, H3K4me2, H3K9me3, etc.), Polycomb-group protein (PRC2: Suz12, PRC1:YY1) and trithorax-group protein (Ash1) to study the epigenetic landscape or RNA Polymerase II to study the transcription landscape.
DamID	Analogously to ChIP, genomic regions bound by a protein of interest can be isolated and used to probe a microarray to determine binding site occupancy. Unlike ChIP, DamID does not require antibodies but makes use of adenine methylation near the protein's binding sites to selectively amplify those regions, introduced by expressing minute amounts of protein of interest fused to bacterial DNA adenine methyltransferase.
SNP detection	Identifying single nucleotide polymorphism among alleles within or between populations. Several applications of microarrays make use of SNP detection, including Genotyping, forensic analysis, measuring predisposition to disease, identifying drug-candidates, evaluating germline mutations in individuals or somatic mutations in cancers, assessing loss of heterozygosity, or genetic linkage analysis.
Alternative splicing	detection An 'exon junction array design uses probes specific to the expected or potential splice sites of predicted exons for a gene. It is of intermediate density, or coverage, to a typical gene expression array (with 1-3 probes per gene) and a genomic tiling array (with hundreds or thousands of probes per gene). It is used to assay the expression of alternative splice forms of a gene. Exon arrays have a different design, employing probes designed to detect each individual exon for known or predicted genes, and can be used for detecting different splicing isoforms.
Fusion genes	A Fusion gene microarray can detect fusion transcripts, e.g. microarray from cancer specimens. The principle behind this is building on the

	alternative splicing microarrays. The oligo design strategy enables combined measurements of chimeric transcript junctions with exon-wise measurements of individual fusion partners.
Tiling array	Genome tiling arrays consist of overlapping probes designed to densely represent a genomic region of interest, sometimes as large as an entire human chromosome. The purpose is to empirically detect expression of transcripts or alternatively splice forms which may not have been previously known or predicted.

Fabrication

Microarrays can be manufactured in different ways, depending on the number of probes under examination, costs, customization requirements, and the type of scientific question being asked. Arrays may have as few as 10 probes or up to 2.1 million micrometre-scale probes from commercial vendors.

Microarrays can be fabricated using a variety of technologies, including printing with fine-pointed pins onto glass slides, photolithography using pre-made masks, photolithography using dynamic micromirror devices, ink-jet printing, or electrochemistry on microelectrode arrays. In spotted microarrays, the probes are oligonucleotides, cDNA or small fragments of PCR products that correspond to mRNAs. The probes are synthesized prior to deposition on the array surface and are then "spotted" onto glass. A common approach utilizes an array of fine pins or needles controlled by a robotic arm that is dipped into wells containing DNA probes and then depositing each probe at designated locations on the array surface. The resulting "grid" of probes represents the nucleic acid profiles of the prepared probes and is ready to receive complementary cDNA or cRNA "targets" derived from experimental or clinical samples. This technique is used by research scientists around the world to produce "in-house" printed microarrays from their own labs. These arrays may be easily customized for each experiment, because researchers can choose the probes and printing locations on the arrays, synthesize the probes in their own lab (or collaborating facility), and spot the arrays. They can then generate their own labeled samples for hybridization, hybridize the samples to the array, and finally scan the arrays with their own equipment. This provides a relatively low-cost microarray that may be customized for each study, and avoids the costs of purchasing often more expensive commercial arrays that may represent vast numbers of genes that are not of interest to the investigator. Publications exist which indicate in-house spotted microarrays may not provide the same level of sensitivity compared to commercial oligonucleotide arrays, possibly owing to the small batch sizes and reduced printing efficiencies when compared to industrial manufactures of oligo arrays.

In oligonucleotide microarrays, the probes are short sequences designed to match parts of the sequence of known or predicted open reading frames. Although oligonucleotide probes are often used in "spotted" microarrays, the term "oligonucleotide array" most often refers to a specific technique of manufacturing. Oligonucleotide arrays are produced by printing short oligonucleotide sequences designed to represent a single gene or family of gene splice-variants by synthesizing this sequence directly onto the array surface instead of depositing intact sequences.

Sequences may be longer (60-mer probes such as the Agilent design) or shorter (25-mer probes produced by Affymetrix) depending on the desired purpose; longer probes are more specific to individual target genes, shorter probes may be spotted in higher density across the array and are cheaper to manufacture. One technique used to produce oligonucleotide arrays include photolithographic synthesis (Affymetrix) on a silica substrate where light and light-sensitive masking agents are used to "build" a sequence one nucleotide at a time across the entire array.

Each applicable probe is selectively "unmasked" prior to bathing the array in a solution of a single nucleotide, then a masking reaction takes place and the next set of probes are unmasked in preparation for a different nucleotide exposure. After many repetitions, the sequences of every probe become fully constructed. More recently, Maskless Array Synthesis from NimbleGen Systems has combined flexibility with large numbers of probes.

Two-channel vs. one-channel detection

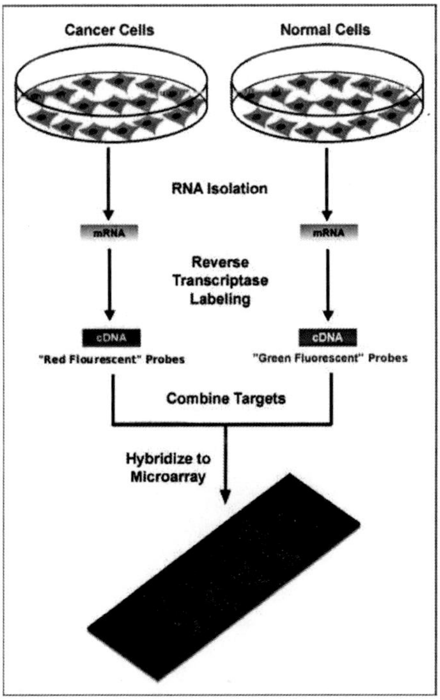

Figure: *Diagram of typical dual-colour microarray experiment.*

Two-color microarrays or two-channel microarrays are typically hybridized with cDNA prepared from two samples to be compared (e.g. diseased tissue versus healthy tissue) and that are labeled with two different fluorophores. Fluorescent dyes commonly used for cDNA labeling include Cy3, which has a fluorescence emission wavelength of 570 nm (corresponding to the green part of the light spectrum), and Cy5 with a fluorescence emission wavelength of 670 nm (corresponding to the red part ofm the light spectrumm). The two Cy-labeled cDNA samples are mixed and hybridized to a single microarray that is then scanned in a microarray scanner to visualize fluorescence of the two fluorophores after excitation with a laser beam of a defined wavelength. Relative intensities of each fluorophore may then be used in ratio-based analysis to identify up-regulated and down-regulated genes.

Oligonucleotide microarrays often carry control probes designed to hybridize with RNA spike-ins. The degree of hybridization between the spike-ins and the control probes is used to normalize the hybridization measurements for the target probes. Although absolute levels of gene expression may be determined in the two-color array in rare instances, the relative differences in expression among different spots within a sample and between samples is the preferred method of data analysis for the two-color system. Examples of providers for such microarrays includes Agilent with their Dual-Mode platform, Eppendorf with their DualChip platform for colorimetric Silverquant labeling, and TeleChem International with Arrayit.

In single-channel microarrays or one-color microarrays, the arrays provide intensity data for each probe or probe set indicating a relative level of hybridization with the labeled target. However, they do not truly indicate abundance levels of a gene but rather relative abundance when compared to other samples or conditions when processed in the same experiment. Each RNA molecule encounters protocol and batch-specific bias during amplification, labeling, and hybridization phases of the experiment making comparisons between genes for the same microarray uninformative. The comparison of two conditions for the same gene requires two separate single-dye hybridizations. Several popular single-channel systems are the Affymetrix "Gene Chip", Illumina "Bead Chip", Agilent single-channel arrays, the Applied Microarrays "CodeLink" arrays, amnd the Eppendorf "DualChip & Silverquant". One strength of the single-dye system lies in the fact that an aberrant sample cannot affect the raw data derived from other samples, because each array chip is exposed to only one sample (as opposed to a two-color system in which a single low-quality sample may drastically impinge on overall data precision even if the other sample was of high quality). Another benefit is that data are more easily compared to arrays from different experiments so long as batch effects have been accounted for. A drawback to the one-color system is that, when compared to the two-color system, twice as many microarrays are needed to compare samples within an experiment.

Microarrays and Bioinformatics

The advent of inexpensive microarray experiments created several specific bioinformatics challenges:

- the multiple levels of replication in experimental design (Experimental design)
- the number of platforms and independent groups and data format (Standardization)
- the treatment of the data (Statistical analysis)
- accuracy and precision (Relation between probe and gene)
- the sheer volume of data and the ability to share it (Data warehousing)

Experimental Design

Due to the biological complexity of gene expression, the considerations of experimental design that are discussed in the expression profiling article are of critical importance if statistically and biologically valid conclusions are to be drawn from the data. There are three main elements to consider when designing a microarray experiment. First, replication of the biological samples is essential for drawing conclusions from the experiment. Second, technical replicates (two RNA samples obtained from each experimental unit) help to ensure precision and allow for testing differences within treatment groups.

The biological replicates include independent RNA extractions and technical replicates may be two aliquots of the same extraction. Third, spots of each cDNA clone or oligonucleotide are present as replicates (at least duplicates) on the microarray slide, to provide a measure of technical precision in each hybridization. It is critical that information about the sample preparation and handling is discussed, in order to help identify the independent units in the experiment and to avoid inflated estimates of statistical significance.

Standardization

Microarray data is difficult to exchange due to the lack of standardization in platform fabrication, assay protocols, and analysis methods. This presents an interoperability problem in bioinformatics. Various grass-roots open-source projects are trying to ease the exchange and analysis of data produced with non-proprietary chips:

- For example, the "Minimum Information About a Microarray Experiment" (MIAME) checklist helps define the level of detail that should exist and is being adopted by many journals as a requirement for the submission of papers incorporating microarray results. But MIAME does not describe the format for the information, so while many formats can support the MIAME requirements, as of 2007 no format permits verification of complete semantic compliance.

- The "MicroArray Quality Control (MAQC) Project" is being conducted by the US Food and Drug Administration (FDA) to develop standards and quality control metrics which will eventually allow the use of MicroArray data in drug discovery, clinical practice and regulatory decision-making.
- The MGED Society has developed standards for the representation of gene expression experiment results and relevant annotations.

Statistical Analysis

Microarray data sets are commonly very large, and analytical precision is influenced by a number of variables. Statistical challenges include taking into account effects of background noise and appropriate normalization of the data.

Normalization methods may be suited to specific platforms and, in the case of commercial platforms, the analysis may be proprietary. Algorithms that affect statistical analysis include:

- Image analysis: gridding, spot recognition of the scanned image (segmentation algorithm), removal or marking of poor-quality and low-intensity features (called flagging).
- Data processing: background subtraction (based on global or local background), determination of spot intensities and intensity ratios, visualisation of data (e.g. see MA plot), and log-transformation of ratios, global or local normalization of intensity ratios, and segmentation into different copy number regions using step detection algorithms.
- Identification of statistically significant changes: t-test, ANOVA, Bayesian method Mann–Whitney test methods tailored to microarray data sets, which take into account multiple comparisons or cluster analysis. These methods assess statistical power based on the variation present in the data and the number of experimental replicates, and can help minimize Type I and type II errors in the analyses.
- Network-based methods: Statistical methods that take the underlying structure of gene networks into account, representing either associative or causative interactions or dependencies among gene products.

Microarray data may require further processing aimed at reducing the dimensionality of the data to aid comprehension and more focused analysis.

Other methods permit analysis of data consisting of a low number of biological or technical replicates; for example, the Local Pooled Error (LPE) test pools standard deviations of genes with similar expression levels in an effort to compensate for insufficient replication.

Relation Between Probe and Gene

The relation between a probe and the mRNA that it is expected to detect is not trivial. Some mRNAs may cross-hybridize probes in the array that are supposed to detect another mRNA. In addition, mRNAs may experience amplification bias that is sequence or molecule-specific. Thirdly, probes that are designed to detect the mRNA of a particular gene may be relying on genomic EST information that is incorrectly associated with that gene.

Data Warehousing

Microarray data was found to be more useful when compared to other similar datasets. The sheer volume of data, specialized formats (such as MIAME), and curation efforts associated with the datasets require specialized databases to store the data.

- Cyanine dyes, such as Cy3 and Cy5, are commonly used fluorophores with microarrays

Cyanine is a non-systematic name of a synthetic dye family belonging to polymethine group. Cyanines have many uses as fluorescent dyes, particularly in biomedical imaging. Depending on the structure, they cover the spectrum from IR to UV. Cyanines were originally used, and still are, to increase the sensitivity range of photographic emulsions, i.e., to increase the range of wavelengths which will form an image on the film. Cyanines are also used in CD-R and DVD-R media. The ones used are mostly green or light blue in color, and are chemically unstable. This makes unstabilized cyanine discs unsuitable for archival CD and DVD use, as they can fade and become unreadable in a few years, however, recent cyanine discs contain stabilizers that slow down the deterioration significantly. These discs are often rated with an archival life of 75 years or more. The other dyes used in CD-Rs are phthalocyanine and azo. Cyanines were first synthesized over a century ago, and there are a large number reported in the literature.

Structure

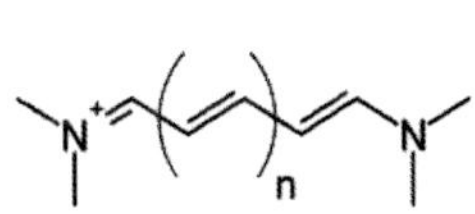

I

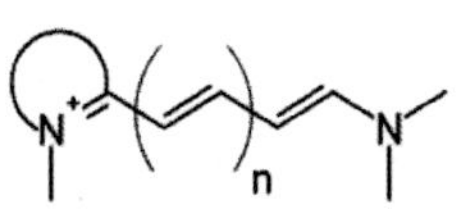

II

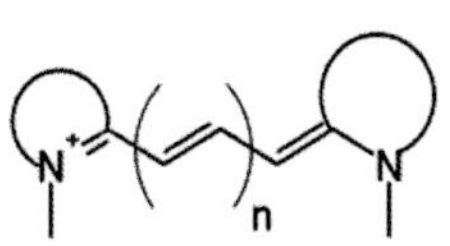

III

Figure:

The Cyanines:

I = Streptocyanines,

II = Hemicyanines,

III = Closed cyanine

There are three types of cyanines:

- Streptocyanines or open chain cyanines:

  $R_2N=CH[CH=CH]_n-NR_2$ (I)

- Hemicyanines:

  $Aryl=N=CH[CH=CH]_n-NR_2$ (II)

- Closed chain cyanines:

  $Aryl=N=CH[CH=CH]_n-N=Aryl$ (III)

where two nitrogens are joined by a polymethine chain. Both nitrogens are each independently part of a heteroaromatic moiety, such as pyrrole, imidazole, thiazole, pyridine, quinoline, indole, benzothiazole, etc.

Cy3 and Cy5

Cy3 and Cy5 are reactive water-soluble fluorescent dyes of the cyanine dye family. Cy3 dyes are red (~550 nm excitation, ~570 nm emission and therefore appear red), while Cy5 is fluorescent in the far red region (~650/670 nm) but absorbs in the orange region (~649 nm). They are usually synthesized with reactive groups on either one or both of the nitrogen side chains so that they can be chemically linked to either nucleic acids or protein molecules. Labeling is done for visualization and quantification purposes. They are used in a wide variety of biological applications including comparative genomic hybridization and in gene chips, which are used in transcriptomics. They are also used to label proteins and nucleic acid for various studies including proteomics and RNA localization.

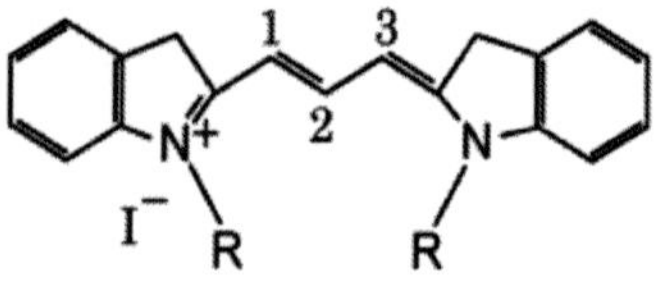

Cy3

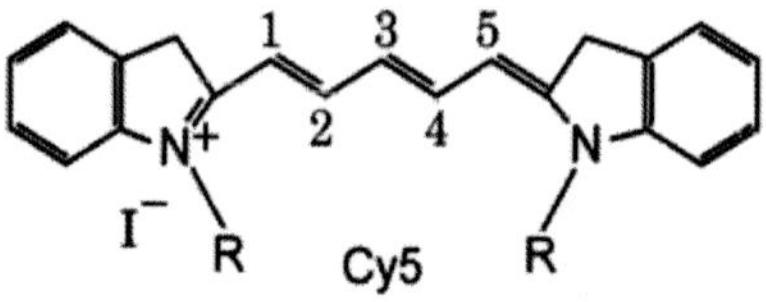

Figure:

Nomenclature and Structure

Standard chemical names specify exactly the chemical structure of the molecule. The Cy3 and Cy5 nomenclature was first proposed by Ernst, et al. in 1989, and is non-standard, since it gives no hint of their chemical structures. In the original paper the number designated the count of the methines (as shown), and the side chains were unspecified. Thus various structures are designated Cy3 and Cy5 in the literature. The R groups do not have to be identical. In the dyes as used they are short aliphatic chains one or both of which ends in a highly reactive moieties such as N-hydroxysuccinimide or maleimide.

Table: *Spectral characteristics*

Dye	*Absorbance Max*	*Emission Max*	*Quantum yield in PBS buffer*	*Molecular weight (Da)*
Cy3	550 nm	570 nm	0.04	766
Cy5	649 nm	670 nm	0.28	792

The scanners actually use different laser emission wavelengths (typically 532 nm and 635 nm) and filter wavelengths (550-600 nm and 655-695 nm) to avoid background contamination. They are thus able to easily distinguish between two samples when one sample has been labeled with Cy3 and the other labeled with Cy5. They are also able to quantify the amount of labeling in either sample.

Cy Dye Alternatives

Alexa Fluor dyes, Atto Dyes, Dylight, IRIS Dyes, Seta dyes, SeTau dyes, SRfluor dyes and Square dyes dyes can be used interchangeably with Cy dyes in most biochemical applications.

Cy5 ozone susceptibility

In 2003, researchers at Inpharmatics and Agilent reported in Analytical Chemistry that microarrays which used Cy5 were susceptible to intermittent data quality decrease caused by environmental ozone. Exposures to ozone levels above 5-10 ppb for 10–30 seconds were reported to decrease the reproducibility of Cy5 microarrays. Much higher levels of ozone (>100 ppb) were required to observe an effect in Cy3.

Applications

Nucleic Acid Labeling

In microarray experiments DNA or RNA is labeled with either Cy3 or Cy5 that has been synthesized to carry an N-hydroxysuccinimidyl ester (NHS-ester) reactive group. Since, NHS-esters react readily only with aliphatic amine groups, which nucleic acids lack, nucleotides have to be modified with aminoallyl groups. This is done through incorporating aminoallyl-modified nucleotides during synthesis reactions. A good ratio is a label every 60 bases such that the labels are not too close to each other, thus resulting in quenching effects.

Protein Labeling

For protein labeling, Cy3 and Cy5 dyes sometimes bear maleimide reactive groups instead. The maleimide functionality allows conjugation of the fluorescent dye to the sulfhydryl group of cysteine residues. Cysteines can be added and removed from the protein domain of interest via PCR mutagenesis. Cy5, is sensitive to the electronic environment it resides in. Changes in the conformation of the protein it is attached to will produce an enhancement or quenching of the emission. The rate of this change can be measured to determine enzyme kinetic parameters. The dyes can be used for similar purposes in FRET experiments. Cy3 and Cy5 are used in proteomics experiments so that samples from two sources can be mixed and run together thorough the separation process. This eliminates variations due to differing experimental conditions that are inevitable if the samples were run separately. These variations make it extremely difficult, if not impossible, to use computers to automate the acquisition of the data after the separation is complete. Using these dyes makes the automation trivial.

Spectrophotometric Quantification of Dye Incorporation

The amount of each sample as well as the amount of incorporated dyes has to be quantified following nucleic acid or protein labeling and purification of the labeling reaction. To determine the dye incorporation rate, the absorbance reading at the wavelength reported for maximum absorbance of the fluorescence dye is used. The corresponding extinction coefficient of the dye is used in the Lambert-Beer Law to determine the dye concentration (c = A / (e * d)). Comparing these values with the DNA, RNA or protein concentration gives a dye incorporation rate (Frequency of incorporation (FOI)).

FOI = C(dye) / C(nucleic acid)

Example: Frequency of Incorporation (FOI) of Cy3 per 1000 bases: FOI(Cy3) = 58.5 * A550/A260

In multiple biological applications only minimal sample amounts can be afforded for these measurements (e.g. array CGH). Therefore, specialized photometers like the NanoPhotometer™ offer the possibility to determine sample concentrations as well as dye incorporation with submicroliter volumes (starting with 0.3 μl). In addition, due to the reduction of the optical pathlength samples are diluted automatically in comparison to standard cuvette measurements. The respective virtual dilution factors are considered by the software of the instrument. Because the measurements are processed with undiluted samples, the reproducibility of the results is very high. And if desired, samples can be retrieved after the measurement for further processing.

Gene Chip Analysis

Microarray technology is a powerful tool for genomic analysis. It gives a global view of the genome in a single experiment. Data analysis of the microarray is a vital part of the experiment. Each microarray study comprises multiple microarrays, each giving tens of thousands of data points. Since the volume of data is growing exponentially as microarrays grow larger, the analysis becomes more challenging. In general the greater the volume of data, the more chances arise for erroneous results.

Handling such large volumes of data requires high end computational infrastructures and programs that can handle multiple data formats. There are already programs available for microarray data analysis on various platforms. However, due to rapid development, diversity in microarray technology, and different data formats, there is always the need for more comprehensive and complete microarray data analysis.

Data Analysis

Data analysis is critical, since any error introduced in the data analysis part will result in biologically insignificant results. In data analysis, the information from the raw data file is further processed to yield meaningful biological results. This part includes data normalization, flagging of the data, averaging the ratio for replicates, clustering of similarly expressed genes, etc. Data has to undergo normalization before further analysis. Normalization removes the non-biological variation between the samples. After the normalization, the ratio is calculated for each gene in the replicate. Based on the ratio, differentially regulated genes are determined. There are various statistical analyses which are also done for confidence analysis. Each replicate data is also examined for various experimental artifacts, bias by computing parameters related to intensity, background, flags, spot details, etc.

Replicates

It is important to note the necessity of replicates in conducting microarray experiments. Like any other quantitative measurement, repeated experiments provide the ability to conduct confidence analysis and identify differentially expressed genes at a given level of confidence. More replicates provide more confidence in determining differentially expressed genes. In practice, three to five replicates would be ideal.

Normalization

Normalization is required to standardize data and focus on biologically relevant changes. There are many sources of systematic variation in microarray experiments that affect the measured gene expression levels such as dye bias, heat and light sensitivity, efficiency of dye incorporation, differences in the labeled cDNA hybridization conditions, scanning conditions, and unequal quantities of starting RNA, etc. Normalization is an important step in adjusting the data set for technical variation and removing relative abundance of gene expression profiles; this is the only point where 1- and 2-color data analyses vary. The normalization method depends on the data. The basic idea behind all the normalization methods is that the expected mean intensity ratio between the two channels should be one. If the observed mean intensity ratio deviates from one, the data is mathematically processed in such a way that the final observed mean intensity ratio becomes one. With the mean intensity ratio adjusted to one, the distribution of the gene expression is centered so that genuine differentials can be identified.

Quality Control

Before analyzing data for biological variation, QC steps must be performed to determine whether the data is fit for statistical testing. Statistical tests are sensitive to the nature of the input data.

Filtering of Flagged Data

Filtering of bad intensity spots is an important process of quality control. For example, the scanner has a measurement limit below which intensity values cannot be trusted. Typically, the lowest intensity value of reliable data is 100–200 for Affymetrix data and 100–1000 for cDNA Microarray data. These cut-offs are likely to change as scanners become more precise. Values below the cut-off point are usually removed (filtered) from the data because they are likely to be artifacts.

Filtering of Noisy Replicates

Filtering of noisy replicates is a crucial part of quality control. Experimental replicates should have similar values. Replicates with noise should be eliminated before analysis; this can be done using the ANOVA statistical method.

Filtering of Non-significant Genes

Filtering of non-significant genes is done so that analysis can be done on selected genes. Non-significant genes are removed by specifying relative change in expression with respect to normal control. Values for over-expressed and under -expressed genes are defined as 2 and “2 respectively. As a result of filtering, few genes are retained. Those remaining genes are then subjected to statistical analysis.

Statistical Analysis

Statistical analysis plays a vital role in identifying genes that are expressed at statistically significant levels.

Clustering

Clustering is a data mining technique used to group genes having similar expression patterns. Hierarchical clustering, and k-mean clustering are widely used techniques in microarray analysis.

Hierarchical Clustering

Hierarchical clustering is a statistical method for finding relatively homogeneous clusters. Hierarchical clustering consists of two separate phases. Initially, a distance matrix containing all the pairwise distances between the genes is calculated. Pearson's correlation and Spearman's correlation are often used as dissimilarity estimates, but other methods, like Manhattan distance or Euclidean distance, can also be applied. If the genes on a single chip are to be clustered, the Euclidean distance is the correct choice, since at least two chips are needed for calculation of any correlation measures. After calculation of the initial distance matrix, the hierarchical clustering algorithm either (A) joins iteratively the two closest clusters starting from single data points (agglomerative, bottom-up approach), or (B) partitions clusters iteratively starting from the complete set (divisive, top-down approach). After each step, a new distance matrix between the newly formed clusters and the other clusters is recalculated. Hierarchical cluster analysis methods include:

- Single linkage (minimum method, nearest neighbor)
- Complete linkage (maximum method, furthest neighbor)
- Average linkage (UPGMA).

K-means Clustering

K-means clustering is an algorithm for classifying and grouping genes based on pattern into K groups. Grouping is done by minimizing the sum of the squares of distances between the data and the corresponding cluster centroid. Thus the purpose of K-mean clustering is to classify data based on similar expression. (www.biostat.ucsf.edu).

Gene Ontology Studies

Gene ontology studies give biologically meaningful information about the gene including cellular location, molecular function, and biological function. This information is analyzed for differences in regulation in disease or drug treatment regimen, with respect to normal control.

Pathway Analysis

Pathway analysis gives specific information about the pathway being affected in disease conditions, with respect to normal control. Pathway analysis also allows identification of gene networks and how genes are regulated.

Tiling Array

Tiling Arrays are a subtype of microarray chips. Like traditional microarrays, they function by hybridizing labeled DNA or RNA target molecules to probes fixed onto a solid surface. Tiling arrays differ from traditional microarrays in the nature of the probes. Instead of probing for sequences of known or predicted genes which may be dispersed throughout the genome, tiling arrays probe intensively for sequences which are known to exist in a contiguous region of genome. This is useful for characterizing regions of genome which are sequenced but with local functions that are largely unknown. Tiling arrays aid in transcriptome mapping as well as in discovering sites of DNA- protein interaction (ChIP-chip), of DNA methylation (MeDIP-chip), and of sensitivity to DNase (DNase Chip), in addition to other uses (eg array CGH). In addition to the advantage of being able to detect previously unidentified genes and regulatory sequences, improved quantification of transcription products is possible. Specific probes are present in millions of copies (as opposed to only several as in traditional arrays) within an array unit called a feature, with anywhere from 10,000 to more than 6,000,000 different features per array. Variable levels of mapping resolution are obtainable by adjusting the amount of sequence overlap between probes, or the amount of known base pairs between probe sequences, as well as the length of the probes themselves. For smaller genomes such as that of Arabidopsis, whole genomes can be examined. Tiling arrays are quickly becoming one of the most powerful tools in genome-wide investigations.

Synthesis and Manufacturers

There are two main ways of synthesizing tiling arrays; the first is photolithographic manufacturing and the second is mechanically spotting or printing. The first method involves in situ synthesis where probes, approximately 25bp, are built on the surface of the chip. These arrays can hold up to 6 million discrete features, each of which contains millions of copies of one probe. The other way of synthesizing tiling array chips is via mechanically printing probes onto the chip. This is done by using automated machines with pins that place the previously synthesized probes onto the surface. Due to the size restriction of the pins, these chips can hold up to nearly 400,000 features. Three manufacturers of tiling arrays are Affymetrix, NimbleGen and Agilent. Their products vary in probe length and spacing.

ChIP-on-chip

ChIP-on-chip (also known as ChIP-chip) is a technique that combines chromatin immunoprecipitation ("ChIP") with microarray technology ("chip"). Like regular ChIP, ChIP-on-chip is used to investigate interactions between proteins and DNA in vivo. Specifically, it allows the identification of the cistrome, sum of binding sites, for DNA-binding proteins on a genome-wide basis. Whole-genome analysis can be performed to determine the locations of binding sites for almost any protein of interest. As the name of the technique suggests, such proteins are generally those operating in the context of chromatin. The most prominent representatives of this class are transcription factors, replication-related proteins, like ORC, histones, their variants, and histone modifications. The goal of ChIP-on-chip is to localize protein binding sites that may help identify functional elements in the genome. For example, in the case of a transcription factor as a protein of interest, one can determine its transcription factor binding sites throughout the genome. Other proteins allow the identification of promoter regions, enhancers, repressors and silencing elements, insulators, boundary elements, and sequences that control DNA replication. If histones are subject of interest, it is believed that the distribution of modifications and their localizations may offer new insights into the mechanisms of regulation. One of the long-term goals ChIP-on-chip was designed for is to establish a catalogue of (selected) organisms that lists all protein-DNA interactions under various physiological conditions. This knowledge would ultimately help in the understanding of the machinery behind gene regulation, cell proliferation, and disease progression. Hence, ChIP-on-chip offers not only huge potential to complement our knowledge about the orchestration of the genome on the nucleotide level, but also on higher levels of information and regulation as it is propagated by research on epigenetics.

Technological Platforms

The technical platforms to conduct ChIP-on-chip experiments are DNA microarrays, or "chips". They can be classified and distinguished according to various characteristics:

- Probe type: DNA arrays can comprise either mechanically spotted cDNAs or PCR-products, mechanically spotted oligonucleotides, or oligonucleotides that are synthesized in situ. The early versions of microarrays were designed to detect RNAs from expressed genomic regions (open reading frames). Although such arrays are perfectly suited to study gene expression profiles, they have limited importance in ChIP experiments since most "interesting" proteins with respect to this technique bind in intergenic regions. Nowadays, even custom-made arrays can be designed and fine-tuned to match the requirements of an experiment. Also, any sequence of nucleotides can be synthesized to cover genic as well as intergenic regions.
- Probe size: Early version of cDNA arrays had a probe length of about 200bp. Latest array versions use oligos as short as 70- (Microarrays, Inc.) to 25-mers (Affymetrix). (Feb 2007)
- Probe composition: There are tiled and non-tiled DNA arrays. Non-tiled arrays use probes selected according to non-spatial criteria, i.e., the DNA sequences used as probes have no fixed distances in the genome. Tiled arrays, however, select a genomic region (or even a whole genome) and divide it into equal chunks. Such a region is called tiled path. The average distance between each pair of neighboring chunks (measured

from the center of each chunk) gives the resolution of the tiled path. A path can be overlapping, end-to-end or spaced.

- Array size: The first microarrays used for ChIP-on-Chip contained about 13,000 spotted DNA segments representing all ORFs and intergenic regions from the yeast genome. Nowadays, Affymetrix offers whole-genome tiled yeast arrays with a resolution of 5bp (all in all 3.2 million probes). Tiled arrays for the human genome become more and more powerful, too. Just to name example, Affymetrix offers a set of seven arrays with about 90 million probes, spanning the complete non-repetitive part of the human genome with about 35bp spacing. (Feb 2007).

Besides the actual microarray, other hard- and software equipment is necessary to run ChIP-on-chip experiments. It is generally the case that one company's microarrays can not be analyzed by another company's processing hardware. Hence, buying an array requires also buying the associated workflow equipment. The most important elements are, among others, hybridization ovens, chip scanners, and software packages for subsequent numerical analysis of the raw data.

Workflow of a ChIP-on-chip Experiment

Starting with a biological question, a ChIP-on-chip experiment can be divided into three major steps: The first is to set up and design the experiment by selecting the appropriate array and probe type. Second, the actual experiment is performed in the wet-lab. Last, during the dry-lab portion of the cycle, gathered data are analyzed to either answer the initial question or lead to new questions so that the cycle can start again.

Wet-lab Portion of the Workflow

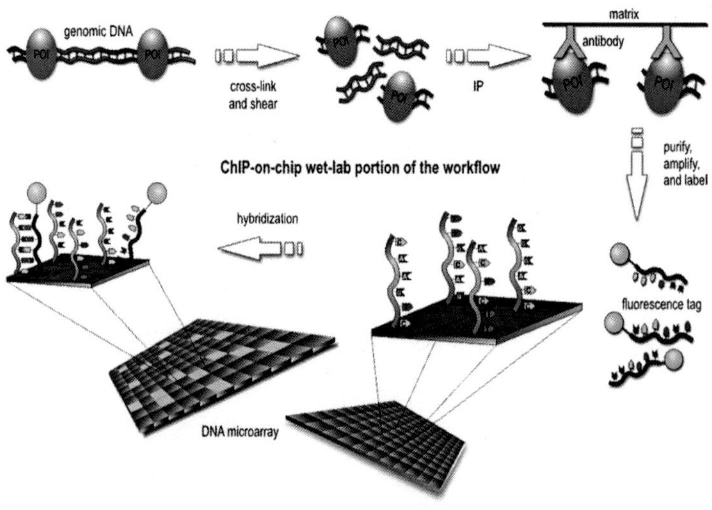

Figure: *Workflow overview of the wet-lab portion of a ChIP-on-chip experiment.*

- In the first step, the protein of interest (POI) is cross-linked with the DNA site it binds to in an in vivo environment. Usually this is done by a gentle formaldehyde fixation that is reversible with heat.
- Then, the cells are lysed and the DNA is sheared by sonication or using micrococcal nuclease. This results in double-stranded chunks of DNA fragments, normally 1 kb or less in length. Those that were cross-linked to the POI form a POI-DNA complex.
- In the next step, only these complexes are filtered out of the set of DNA fragments, using an antibody specific to the POI. The antibodies may be attached to a solid surface, may have a magnetic bead, or some other physical property that allow distributing cross-linked complexes and unbound fragments. This procedure is essentially an immunoprecipitation (IP). There are two alternative ways to implement this filtering step:
    - o immunoprecipitation of the tagged protein with an antibody against the tag (ex. FLAG, HA, c-myc)
    - o affinity purification that does not require antibodies, such as the Tandem Affinity Purification (TAP)
- The cross-linking of POI-DNA complexes is reversed (usually by heating) and the DNA strands are purified. For the rest of the workflow, the POI is no longer necessary.
- After an amplification and denaturation step, the single-stranded DNA fragments are labeled with a fluorescent tag such as Cy5 or Alexa 647.
- Finally, the fragments are poured over the surface of the DNA microarray, which is spotted with short, single-stranded sequences that cover the genomic portion of interest. Whenever a labeled fragment "finds" a complementary fragment on the array, they will hybridize and form again a double-stranded DNA fragment.
- Quantitative and qualitative DNA and protein analysis with very small volumes of valuable samples are essential steps during the whole wet-lab workflow. Specialized NanoPhotometer offer the possibility of UV-visible spectrophotometry with sample volumes of 0.3 μl.

  Reduction of the optical pathlength, compared to standard measurements, effects an automatic dilution of the samples. Therefore, samples can be measured undiluted and if desired, retrieved after the measurement for further processing.

Dry-lab Portion of the Workflow

- After a sufficiently large time frame to allow hybridization, the array is illuminated with fluorescence light. Those probes on the array that are hybridized to one of the labeled fragments emit a light signal that is captured by a camera. This image contains all raw data for the remaining part of the workflow.

- This raw data, encoded as false-color image, needs to be converted to numerical values before the actual analysis can be done. The analysis and information extraction of the raw data often remains the most challenging part for ChIP-on-chip experiments. Problems arise throughout this portion of the workflow, ranging from the initial chip read-out, to suitable methods to subtract background noise, and finally to appropriate algorithms that normalize the data and make it available for subsequent statistical analysis, which then hopefully lead to a better understanding of the biological question sought to answer. Furthermore, due to the different array platforms and missing standardization between them, data storage and exchange is a huge problem, too. Generally speaking, the data analysis can be divided into three major steps:
  - o During the first step, the captured fluorescence signals from the array are normalized, using control signals derived from the same or a second chip. Such control signals tell which probes on the array were hybridized correctly and which bound nonspecifically.
  - o In the second step, numerical and statistical tests are applied to control data and IP fraction data to identify POI-enriched regions along the genome. The following three methods are used widely: Median percentile rank, Single-array error, and Sliding-window. These methods generally differ in a way how low-intensity signals are handled, how much background noise is accepted, and which trait for the data is emphasized during the computation. In the recent past, the sliding-window approach seems to be favored and is often described as most powerful.
  - o In the third step, these regions are analyzed further. If, for example, the POI was a transcription factor, such regions would represent its binding sites. Subsequent analysis then may want to infer nucleotide motifs and other patterns to allow functional annotation of the genome.

Strengths and Weaknesses

Using tiled arrays, ChIP-on-chip allows for high resolution of genome-wide maps. These maps can determine the binding sites of many DNA-binding proteins like transcription factors and also chromatin modifications.

Although ChIP-on-chip can be a powerful technique in the area of genomics, it is very expensive. Most published studies using ChIP-on-chip repeat their experiments at least three times to ensure biologically meaningful maps. The cost of the DNA microarrays is often a limiting factor to whether a laboratory should proceed with a ChIP-on-chip experiment. Another limitation is the size of DNA fragments that can be achieved.

Most ChIP-on-chip protocols utilize sonication as a method of breaking up DNA into small pieces. However, sonication is limited to a minimal fragment size of 200 bp. For higher resolution maps, this limitation should be overcome to achieve smaller fragments, preferably to single nucleosome resolution. As mentioned previously, the statistical analysis of the huge amount of data generated from arrays is a challenge and normalization procedures should aim to minimize artifacts and determine what is really biologically significant. So far, application to mammalian genomes has been a major limitation, for example, due to a significant percentage of the genome that is occupied by repeats.

However, as ChIP-on-chip technology advances, high resolution whole mammalian genome maps should become achievable. Antibodies used for ChIP-on-chip can be an important limiting factor. ChIP-on-chip requires highly specific antibodies that must recognize its epitope in free solution and also under fixed conditions. If it is demonstrated to successfully immunoprecipitate cross-linked chromatin, it is termed "ChIP-grade".

Companies that provide ChIP-grade antibodies include Abcam, Cell Signaling Technology, Santa Cruz, and Upstate. To overcome the problem of specificity, the protein of interest can be fused to a tag like FLAG or HA that are recognized by antibodies. An alternative to ChIP-on-chip that does not require antibodies is DamID. Also available are antibodies against a specific histone modification like H3 tri methyl K4. As mentioned before, the combination of these antibodies and ChIP-on-chip has become extremely powerful in determining whole genome analysis of histone modification patterns and will contribute tremendously to our understanding of the histone code and epigenetics.A study demonstrating the non-specific nature of DNA binding proteins has been published in PLoS Biology. This indicates that alternate confirmation of functional relevancy is a necessary step in any ChIP-chip experiment.

History

A first ChIP-on-chip experiment was performed in 1999 to analyze the distribution of cohesin along budding yeast chromosome III. Although the genome was not completely represented, the protocol in this study remains equivalent as those used in later studies. The ChIP-on-chip technique using all of the ORFs of the genome (that nevertheless remains incomplete, missing intergenic regions) was then applied successfully in three papers published in 2000 and 2001. The authors identified binding sites for individual transcription factors in the budding yeast Saccharomyces cerevisiae. In 2002, Richard Young's group determined the genome-wide positions of 106 transcription factors using a c-Myc tagging system in yeast. Other applications for ChIP-on-chip include DNA replication, recombination, and chromatin structure. Since then, ChIP-on-chip has become a powerful tool in determining genome-wide maps of histone modifications and many more transcription factors. ChIP-on-chip in mammalian systems has been difficult due to the large and repetitive genomes. Thus, many studies in mammalian cells have focused on select promoter regions that are predicted to bind transcription factors and have not analyzed the entire genome. However, whole mammalian genome arrays have recently become commercially available from companies like Nimblegen. In the future, as ChIP-on-chip arrays become more and more advanced, high resolution whole genome maps of DNA-binding proteins and chromatin components for mammals will be analyzed in more detail.

ChIP-chip

ChIP-chip is one of the most popular usages of tiling arrays. Chromatin immunoprecipitation is the technique whereby binding sites of proteins can be identified. A genome-wide variation of this is known as ChIP-on-chip. Proteins that bind to chromatin are cross-linked in vivo, usually via fixation with formaldehyde. The chromatin is then fragmented and exposed to antibodies specific to protein of interest. These complexes are then precipitated. The DNA is then isolated and purified.

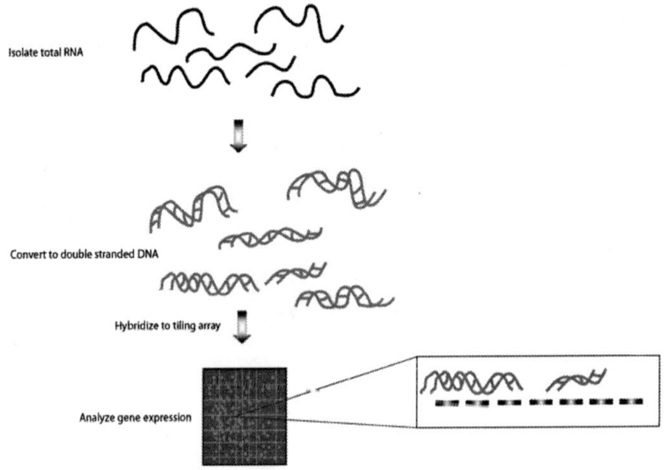

Figure: *Overview of transcriptome mapping procedure.*

With traditional DNA microarrays, the immunoprecipitated DNA is hybridized to the chip, which contains probes, designed to cover regions representative of the genome. However, with tiling arrays, overlapping probes or probes in very close proximity can be used. This gives an unbiased analysis with high resolution. Besides these advantages, tiling arrays show high reproducibility, and with overlapping probes spanning large segments of the genome, tiling arrays can still interrogate protein binding sites, which harbor repeats. ChIP-chip experiments have been done to identify binding sites of transcription factors across the genome in yeast, drosophila and a few mammalian species.

Transcriptome mapping

Another popular use of tiling arrays is in finding expressed genes. Traditional methods of gene prediction for annotation of genomic sequences have had several problems when used to map the transcriptome, such as not producing an accurate structure of the genes and also the missing of transcripts. The method of sequencing cDNA to find transcribed genes also runs into problems, like not being able to detect rare RNA molecules, RNA that are not polyadenylated, and very short RNA, and so would not detect genes that are only active in response to signals or specific to a time frame. Tiling arrays can solve these issues in mapping the transcriptome. Due to the high resolution and sensitivity, even small and rare molecules can be detected.

The overlapping nature of the probes also allows detection of non-polyadenylated RNA and can produce a more precise picture of gene structure. Earlier studies done on chromosome 21 and 22 showed the power of tiling arrays for identifying transcription units. The authors used 25-mer probes that were 35bp apart, spanning the entire chromosomes. Labeled targets were made from polyadenylated RNA. They found many more transcripts than predicted and 90% were outside of annotated exons. Another study done in Arabidopsis used high-density oligonucleotide arrays that cover the entire genome. More than 10 times more transcripts were found than predicted by ESTs and other prediction tools. Also found were novel transcripts in the centromeric regions where it was thought that no genes are actively expressed. Many noncoding and natural antisense RNA have also been identified using tiling arrays.

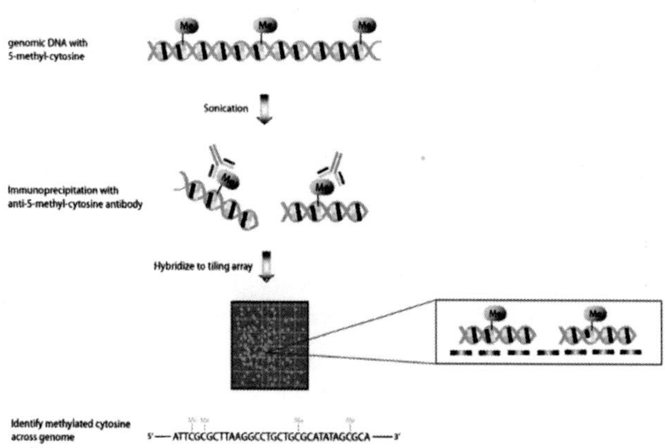

Figure: *Overview of MeDIP-chip procedure.*

MeDIP-chip

Methyl-DNA immunoprecipitation followed by tiling array allows DNA methylation mapping and measurement across the genome. DNA is methylated on cytosine in CG di-nucleotides in many places in the genome. This modification is one of the best-understood inherited epigenetic changes and is shown to affect gene expression. Mapping these sites can add to the knowledge of expressed genes and also epigenetic regulation on a genome-wide level. Studies have been done, utilizing tiling arrays, to generate high-resolution methylation maps for the Arabidopsis genome to generate the first "methylome".

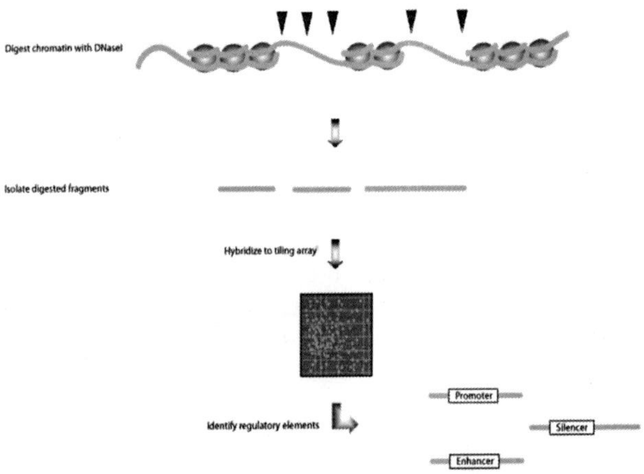

Figure: *Overview of DNase-chip procedure.*

DNase-chip

DNase chip is an application of tiling arrays to identify hypersensitive sites, which are segments of open chromatin that are more readily cleaved by DNaseI. DNaseI cleaving produces larger fragments of around 1.2kb in size. These hypersensitive sites have been shown to be an accurate way of predicting regulatory elements such as promoter regions, enhancers and silencers. Historically, the method uses Southern blotting to find digested fragments; however, tiling arrays have been used in its place for applying the technique to a genome-wide scale.

Comparative Genomic Hybridization (CGH)

Array-based CGH is a technique often used in diagnostics to compare differences between types of DNA, such as normal cells vs. cancer cells. There are two types of tiling arrays commonly used for array CGH, which are the whole genome and fine tiled. The whole genome approach would be useful in identifying copy number variations with high resolution. On the other hand, the fine tiled array CGH would produce ultrahigh resolution to find other abnormalities such as breakpoints.

Procedure

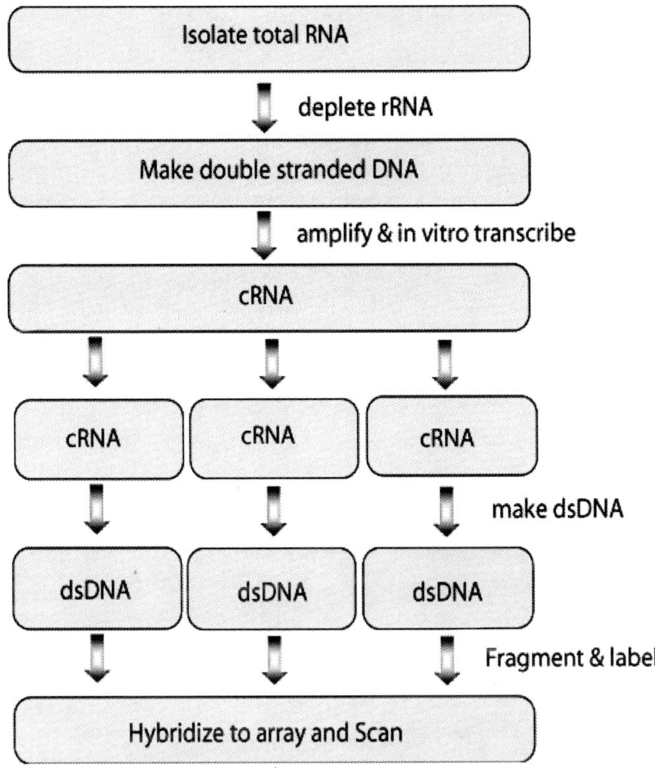

***Figure:** Workflow of tiling array procedure.*

There are several different methods for conducting a tiling arraying. One protocol for analyzing gene expression involves first isolating total RNA. This is then purified of rRNA molecules. The RNA is copied into double stranded DNA, which is subsequently amplified and in vitro transcribed to cRNA. The product is split into triplicates to produce dsDNA, which is then fragmented and labeled. Finally, the samples are hybridized to the tiling array chip. The signals from the chip are scanned and interpreted by computers.

Various software and algorithms are available for data analysis and vary in benefits depending on the manufacturer of the array chip. For Affymetrix chips, the model-based analysis of tiling array (MAT) is the most effective peak-seeking algorithm. For NimbleGen chips, TAMAL is more suitable for locating binding sites. Alternative algorithms include MA2C and TileScope, which are less complicated to operate. The Joint binding deconvolution algorithm is commonly used for the Agilent chips. If sequence analysis of binding site or annotation of the genome is required then programs like MEME, Gibbs Motif Sampler, Cis-regulatory element annotation system and Galaxy are used.

Advantages and Disadvantages

The obvious advantages of tiling array are that they provide an unbiased tool to investigate protein binding, gene expression and gene structure on a genome-wide scope. They allow a new level of insight in studying the transcriptome and methylome.

However, there are also certain drawbacks. First and foremost is the issue of expense. Although the cost of purchasing tiling array kits have reduced in price in the last several years, at the moment, the price makes it impractical to actually use genome wide tiling arrays for larger genomes like mammalian. Another issue is to do with the ultra sensitive detection of this technology. For looking at gene expression, an argument against a study in Arabidopsis sp., which found ten times more genes in the genome than traditional prediction tools, is that the results were confounded by "transcriptional noise". Furthermore there is an analysis challenge, as there is no clearly defined start or stop to your region of interest once identified on the array. Also, the arrays usually only give chromosome and position numbers, necessitating the need to sequence your region of interest (if required) as a separate step (although some modern arrays also give sequence information).

RNA-Seq

RNA-seq, also called "Whole Transcriptome Shotgun Sequencing" ("WTSS") and dubbed "a revolutionary tool for transcriptomics", refers to the use of high-throughput sequencing technologies to sequence cDNA in order to get information about a sample's RNA content, a technique that is quickly becoming invaluable in the study of diseases like cancer. Thanks to the deep coverage and base level resolution provided by next-generation sequencing instruments, RNA-seq provides researchers with efficient ways to measure transcriptome data experimentally, allowing them to get information such as how different alleles of a gene are expressed, detect post-transcriptional mutations or identify gene fusions.

Introduction

The introduction of next-generation sequencing or high-throughput sequencing technologies opened new doors into the field of DNA sequencing, however as understanding of these technologies becomes more widespread and new tools are developed, new innovative ways of applying these technologies are being created. Given high-throughput sequencing technologies' low requirements of nucleotide sequence product, together with its deep coverage and base-scale resolution, its use has expanded to the field of transcriptomics. Transcriptomics is an area of research characterizing the RNA transcribed from a particular genome under investigation. Although transcriptomes are more dynamic than genomic DNA, these molecules provide direct access to gene regulation and protein information. Sequencing transcriptomes is not a new idea. Various methods have been developed previously to directly determine cDNA sequences based mostly around traditional (and more expensive) Sanger sequencing, while others include methodologies such as Serial analysis of gene expression (SAGE), cap analysis gene expression (CAGE) and massively parallel signature sequencing (MPSS). Transcriptome Sequencing (RNA-seq) can be done with a variety of platforms to test a high amount of ideas and hypotheses. For example, using the Illumina (company) Genome Analyzer platform, recent applications include sequencing mammalian transcriptomes, ABI Solid Sequencing to profile stem cell transcriptomes or Life Science's 454 Sequencing to discover SNPs in maize. Even though each platform has its technical differences, the information gathered from each is of the same nature.

Methods

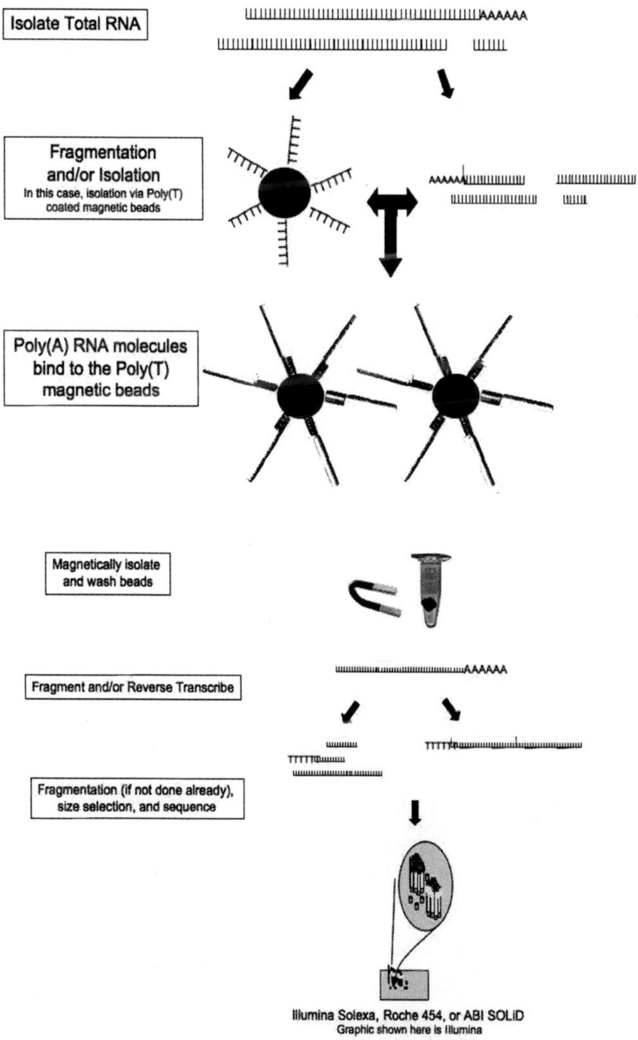

RNA Poly(A) Library

Creation of a sequence library can change from platform to platform in high throughput sequencing, where each has several kits designed to build different types of libraries and adapting the resulting sequences to the specific requirements of their instruments. However, due to the nature of the template being analyzed, there are commonalities within each technology. Frequently, in mRNA analysis the 3' polyadenylated (poly(A)) tail is targeted in order to ensure that coding RNA is separated from noncoding RNA. This can be accomplished simply with poly (T) oligos covalently attached to a given substrate. Presently many studies utilize magnetic beads for this step (;). The Protocol Online website provides a list of several protocols relating to mRNA isolation.

Studies including portions of the transcriptome outside poly(A) RNAs have shown that when using poly(T) magnetic beads, the flow-through RNA (non-poly(A) RNA) can yield important noncoding RNA gene discovery which would have otherwise gone unnoticed. Also, since ribosomal RNA represents over 90% of the RNA within a given cell, studies have shown that its removal via probe hybridization increases the capacity to retrieve data from the remaining portion of the transcriptome.

The next step is reverse transcription. Due to the 5' bias of randomly primed-reverse transcription as well as secondary structures influencing primer binding sites, hydrolysis of RNA into 200-300 nucleotides prior to reverse transcription reduces both problems simultaneously.

However, there are trade-offs with this method where although the overall body of the transcripts are efficiently converted to DNA, the 5' and 3' ends are less so. Depending on the aim of the study, researchers may choose to apply or ignore this step. Once the cDNA is synthesized it can be further fragmented to reach the desired fragment length as specified in table 1. The template is now ready to be prepared for the desired sequencing method.

Polyadenylation

Polyadenylation is the addition of a poly(A) tail to an RNA molecule. The poly(A) tail consists of multiple adenosine monophosphates; in other words, it is a stretch of RNA that has only adenine bases. In eukaryotes, polyadenylation is part of the process that produces mature messenger RNA (mRNA) for translation. It, therefore, forms part of the larger process of gene expression.

The process of polyadenylation begins as the transcription of a gene finishes. The 3'-most segment of the newly-made RNA is first cleaved off by a set of proteins; these proteins then synthesize the poly(A) tail at the RNA's 3' end. In some genes, these proteins may add a poly(A) tail at any one of several possible sites. Therefore, polyadenylation can produce more than one transcript from a single gene, similar to alternative splicing.

The poly(A) tail is important for the nuclear export, translation, and stability of mRNA. The tail is shortened over time, and, when it is short enough, the mRNA is enzymatically degraded. However, in a few cell types, mRNAs with short poly(A) tails are stored for later activation by re-polyadenylation in the cytosol.

In contrast, when polyadenylation occurs in bacteria, it promotes RNA degradation. This is also sometimes the case for eukaryotic non-coding RNAs. The wide distribution of polyadenylation among living organisms indicates that this process evolved early in the history of life on Earth.

Primer on RNA

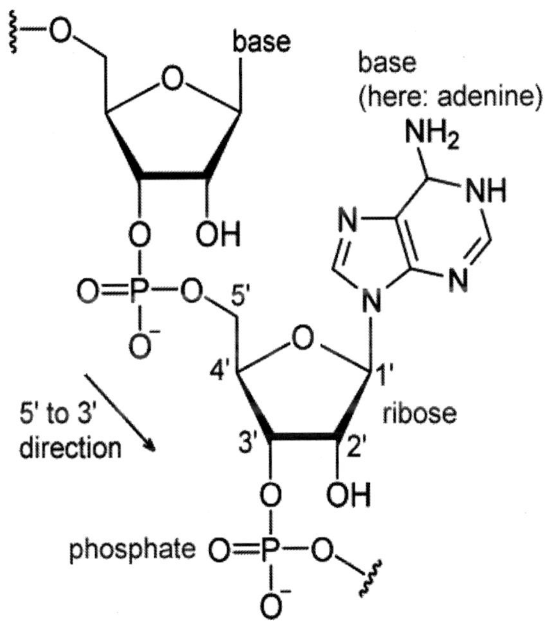

Figure: *Chemical structure of RNA. The sequence of bases differs between RNA molecules.*

RNAs are a type of large biological molecules, whose individual building blocks are called nucleotides. The name poly(A) tail (for polyadenylic acid tail) reflects the way RNA nucleotides are abbreviated, with a letter for the base the nucleotide contains (A for adenine, C for cytosine, G for guanine and U for uracil). RNAs are produced (transcribed) from a DNA template. By convention, RNA sequences are written in a 5' to 3' direction. The 5' end is the part of the RNA molecule that is transcribed first, and the 3' end is transcribed last. The 3' end is also where the poly(A) tail is found on polyadenylated RNAs. Messenger RNA (mRNA) is RNA that has a coding region that acts as a template for protein synthesis (translation). The rest of the mRNA, the untranslated regions, tune how active the mRNA is. There are also many RNAs that are not translated, called non-coding RNAs. Like the untranslated regions, many of these non-coding RNAs have regulatory roles.

Nuclear Polyadenylation

Function

In nuclear polyadenylation, a poly(A) tail is added to an RNA at the end of transcription. On mRNAs, the poly(A) tail protects the mRNA molecule from enzymatic degradation in the cytoplasm and aids in transcription termination, export of the mRNA from the nucleus, and translation. Almost all eukaryotic mRNAs are polyadenylated, with the exception of animal replication-dependent histone mRNAs. These are the only mRNAs in eukaryotes that lack a poly(A) tail, ending instead in a stem-loop structure followed by a purine-rich sequence, termed histone downstream element, that directs where the RNA is cut so that the 3' end of the histone mRNA is formed.

Many eukaryotic non-coding RNAs are always polyadenylated at the end of transcription. There are small RNAs where the poly(A) tail is seen only in intermediary forms and not in the mature RNA as the ends are removed during processing, the notable ones being microRNAs. But, for many long noncoding RNAs – a seemingly large group of regulatory RNAs that, for example, includes the RNA Xist, which mediates X chromosome inactivation – a poly(A) tail is part of the mature RNA.

Mechanism

The polyadenylation machinery in the nucleus of eukaryotes works on products of RNA polymerase II, such as precursor mRNA. Here, a multi-protein complex (see components on the right) cleaves the 3'-most part of a newly produced RNA and polyadenylates the end produced by this cleavage. The cleavage is catalysed by the enzyme CPSF and occurs 10–30 nucleotides downstream of its binding site. This site is often the sequence AAUAAA on the RNA, but variants of it that bind more weakly to CPSF exist. Two other proteins add specificity to the binding to an RNA: CstF and CFI. CstF binds to a GU-rich region further downstream of CPSF's site. CFI recognises a third site on the RNA (a set of UGUAA sequences in mammals) and can recruit CPSF even if the AAUAAA sequence is missing. The polyadenylation signal – the sequence motif recognised by the RNA cleavage complex – varies between groups of eukaryotes. Most human polyadenylation sites contain the AAUAAA sequence, but this sequence is less common in plants and fungi.

The RNA is cleaved right after transcription, as CstF also binds to RNA polymerase II. Cleavage also involves the protein CFII, though it is unknown how. The cleavage site associated with a polyadenylation signal can vary up to some 50 nucleotides.

When the RNA is cleaved, polyadenylation starts, catalysed by polyadenylate polymerase. Polyadenylate polymerase builds the poly(A) tail by adding adenosine monophosphate units from adenosine triphosphate to the RNA, cleaving off pyrophosphate. Another protein, PAB2, binds to the new, short poly(A) tail and increases the affinity of polyadenylate polymerase for the RNA. When the poly(A) tail is approximately 250 nucleotides long the enzyme can no longer bind to CPSF and polyadenylation stops, thus determining the length of the poly(A) tail. CPSF is in contact with RNA polymerase II, allowing it to tell the polymerase to terminate transcription. The polyadenylation machinery is also physically linked to the spliceosome, a complex that removes introns from RNAs.

Downstream Effects

The poly(A) tail acts as the binding site for poly(A)-binding protein. Poly(A)-binding protein promotes export from the nucleus and translation, and inhibits degradation. This protein binds to the poly(A) tail prior to mRNA export from the nucleus and in yeast also recruits poly(A) nuclease, an enzyme that shortens the poly(A) tail and allows the export of the mRNA. Poly(A)-binding protein is exported to the cytoplasm with the RNA. mRNAs that are not exported are degraded by the exosome. Poly(A)-binding protein also can bind to, and thus recruit, several proteins that affect translation, one of these is initiation factor-4G, which in turn recruits the 40S ribosomal subunit. However, a poly(A) tail is not required for the translation of all mRNAs.

Deadenylation

In eukaryotic somatic cells, the poly(A) tail of most mRNAs in the cytoplasm gradually get shorter, and mRNAs with shorter poly(A) tail are translated less and degraded sooner. However, it can take many hours before an mRNA is degraded. This deadenylation and degradation process can be accelerated by microRNAs complementary to the 3' untranslated region of an mRNA. In immature egg cells, mRNAs with shortened poly(A) tails are not degraded, but are instead stored without being translated. They are then activated by cytoplasmic polyadenylation after fertilisation, during egg activation.

In animals, poly(A) ribonuclease (PARN) can bind to the 5' cap and remove nucleotides from the poly(A) tail. The level of access to the 5' cap and poly(A) tail is important in controlling how soon the mRNA is degraded. PARN deadenylates less if the RNA is bound by the initiation factors 4E (at the 5' cap) and 4G (at the poly(A) tail), which is why translation reduces deadenylation. The rate of deadenylation may also be regulated by RNA-binding proteins. Once the poly(A) tail is removed, the decapping complex removes the 5' cap, leading to a degradation of the RNA. Several other enzymes that seem to be involved in deadenylation have been identified in yeast.

Alternative Polyadenylation

Many protein-coding genes have more than one polyadenylation site, so a gene can code for several mRNAs that differ in their 3' end. Since alternative polyadenylation changes the length of the 3' untranslated region, it can change which binding sites for microRNAs the 3' untranslated region contains. MicroRNAs tend to repress translation and promote degradation of the mRNAs they bind to, although there are examples of microRNAs that stabilise transcripts. Alternative polyadenylation can also shorten the coding region, thus making the mRNA code for a different protein, but this is much less common than just shortening the 3' untranslated region.

The choice of poly(A) site can be influenced by extracellular stimuli and depends on the expression of the proteins that take part in polyadenylation. For example, the expression of CstF-64, a subunit of cleavage stimulatory factor (CstF), increases in macrophages in response to lipopolysaccharides (a group of bacterial compounds that trigger an immune response). This results in the selection of weak poly(A) sites and thus shorter transcripts. This removes regulatory elements in the 3' untranslated regions of mRNAs for defense-related products like lysozyme and TNF-á. These mRNAs then have longer half-lives and produce more of these proteins. RNA-binding proteins other than those in the polyadenylation machinery can also affect whether a polyadenyation site is used,as can DNA methylation near the polyadenylation signal.

Cytoplasmic Polyadenylation

There is polyadenylation in the cytosol of some animal cell types, namely in the germ line, during early embryogenesis and in post-synaptic sites of nerve cells. This lengthens the poly(A) tail of an mRNA with a shortened poly(A) tail, so that the mRNA will be translated. These shortened poly(A) tails are often less than 20 nucleotides, and are lengthened to around 80–150 nucleotides.

In the early mouse embryo, cytoplasmic polyadenylation of maternal RNAs from the egg cell allows the cell to survive and grow even though transcription does not start until the middle of the 2-cell stage (4-cell stage in human). In the brain, cytoplasmic polyadenylation is active during learning and could play a role in long-term potentiation, which is the strengthening of the signal transmission from a nerve cell to another in response to nerve impulses and is important for learning and memory formation.

Cytoplasmic polyadenylation requires the RNA-binding proteins CPSF and CPEB, and can involve other RNA-binding proteins like Pumilio. Depending on the cell type, the polymerase can be the same type of polyadenylate polymerase (PAP) that is used in the nuclear process, or the cytoplasmic polymerase GLD-2.

Tagging for Degradation in Eukaryotes

For many non-coding RNAs, including tRNA, rRNA, snRNA, and snoRNA, polyadenylation is a way of marking the RNA for degradation, in at least yeast. This polyadenylation is done in the nucleus by the TRAMP complex, which adds a tail that is around 40 nucleotides long to the 3' end. The RNA is then degraded by the exosome. Poly(A) tails have also been found on human rRNA fragments, both the form of homopolymeric (A only) and heterpolymeric (mostly A) tails.

In Prokaryotes and Organelles

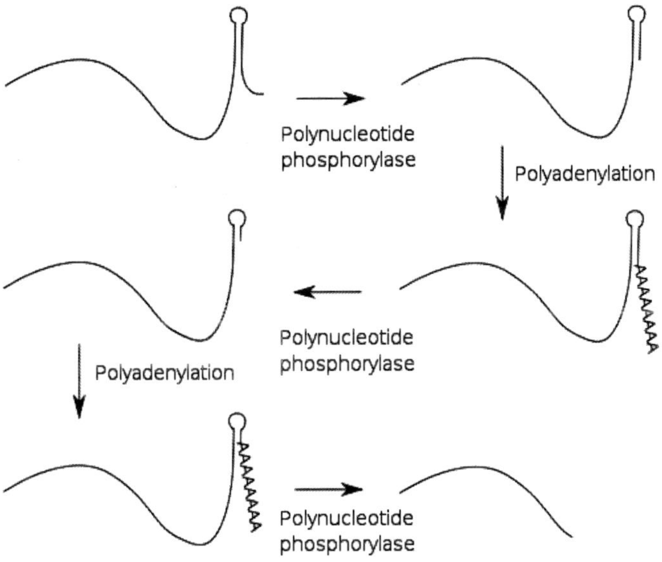

Figure: *Polyadenylation in bacteria helps polynucleotide phosphorylase degrade past secondary structure.*

In many bacteria, both mRNAs and non-coding RNAs can be polyadenylated. This poly(A) tail promotes degradation by the degradosome, which contains two RNA-degrading enzymes: polynucleotide phosphorylase and RNase E. Polynucleotide phosphorylase binds to the 3' end of RNAs and the 3' extension provided by the poly(A) tail allows it to bind to the RNAs whose secondary structure would otherwise block the 3' end. Successive rounds of polyadenylation and degradation of the 3' end by polynucleotide phosphorylase allows the degradosome to overcome these secondary structures. The poly(A) tail can also recruit RNases that cut the RNA in two. These bacterial poly(A) tails are about 30 nucleotides long.

In as different groups as animals and trypanosomes, the mitochondria contain both stabilising and destabilising poly(A) tails. Destabilising polyadenylation targets both mRNA and noncoding RNAs. The poly(A) tails are 43 nucleotides long on average. The stabilising ones start at the stop codon, and without them the stop codon (UAA) is not complete as the genome only encodes the U or UA part. Plant mitochondria have only destabilising polyadenylation, and yeast mitochondria have no polyadenylation at all. While many bacteria and mitochondria have polyadenylate polymerases, they also have another type of polyadenylation, performed by polynucleotide phosphorylase itself. This enzyme is found in bacteria, mitochondria, plastids and as a constituent of the archeal exosome (in those archaea that have an exosome). It can synthesise a 3' extension where the vast majority of the bases are adenines. Like in bacteria, polyadenylation by polynucleotide phosphorylase promotes degradation of the RNA in plastids and likely also archaea.

Evolution

Although polyadenylation is seen in almost all organisms, it is not universal. However, the wide distribution of this modification and the fact that it is present in organisms from all three domains of life implies that the last universal common ancestor of all living organisms, it is presumed, had some form of polyadenylation system. A few organisms do not polyadenylate mRNA, which implies that they have lost their polyadenylation machineries during evolution. Although no examples of eukaryotes that lack polyadenylation are known, mRNAs from the bacterium Mycoplasma gallisepticum and the salt-tolerant archaean Haloferax volcanii lack this modification.

The most ancient polyadenylating enzyme is polynucleotide phosphorylase. This enzyme is part of both the bacterial degradosome and the archaeal exosome, two closely related complexes that recycle RNA into nucleotides. This enzyme degrades RNA by attacking the bond between the 3'-most nucleotides with a phosphate, breaking off a diphosphate nucleotide. This reaction is reversible, and so the enzyme can also extend RNA with more nucleotides. The heteropolymeric tail added by polynucleotide phosphorylase is very rich in adenine. The choice of adenine is most likely the result of higher ADP concentrations than other nucleotides as a result of using ATP as an energy currency, making it more likely to be incorporated in this tail in early lifeforms. It has been suggested that the involvement of adenine-rich tails in RNA degradation prompted the later evolution of polyadenylate polymerases (the enzymes that produce poly(A) tails with no other nucleotides in them).

Polyadenylate polymerases are not as ancient. They have separately evolved in both bacteria and eukaryotes from CCA-adding enzyme, which is the enzyme that completes the 3' ends of tRNAs. Its catalytic domain is homologous to that of other polymerases. It is presumed that the horizontal transfer of bacterial CCA-adding enzyme to eukaryotes allowed the archaeal-like CCA-adding enzyme to switch function to a poly(A) polymerase. Some lineages, like archaea and cyanobacteria, never evolved a polyadenylate polymerase.

transcripts (such as from housekeeping genes) are responsible for the majority of the sequencing data. In a typical sample, 5% of the genes give rise to 75% of the reads sequenced. As a consequence, it is hard to measure the abundances of the remaining genes reliably, and the majority of transcript measurements is very noisy.

Coverage as Measure of Expression

Expression can be deduced via RNA-seq to the extent at which a sequence is retrieved. Transcriptome studies in yeast show that in this experimental setting, a fourfold coverage is required for amplicons to be classified and characterized as an expressed gene. When the transcriptome is fragmented prior to cDNA synthesis, the number of reads corresponding to the particular exon normalized by its length in vivo yields gene expression levels which correlate with those obtained through qPCR. Single nucleotide polymorphism

A single-nucleotide polymorphism (SNP, pronounced snip) is a DNA sequence variation occurring when a single nucleotide — A, T, C or G — in the genome (or other shared sequence) differs between members of a biological species or paired chromosomes in an individual. For example, two sequenced DNA fragments from different individuals, AAGCCTA to AAGCTTA, contain a difference in a single nucleotide. In this case we say that there are two alleles: C and T. Almost all common SNPs have only two alleles. SNPs can occur in both coding and non-coding regions of genome. Within a population, SNPs can be assigned a minor allele frequency — the lowest allele frequency at a locus that is observed in a particular population. This is simply the lesser of the two allele frequencies for single-nucleotide polymorphisms. There are variations between human populations, so a SNP allele that is common in one geographical or ethnic group may be much rarer in another. These genetic variations between the individuals (particularly in the non-coding parts of genome) are exploited in DNA fingerprinting, which is used in criminology. Also, these genetic variations underlie differences in our susceptibility to, or protection from all kinds of diseases. The severity of illness and the way our body responds to treatments are also manifestations of genetic variations. For example, a single base difference in the Apolipoprotein E is associated with Alzheimer's disease.

Types

Types of SNPs

- Non-coding region
- Coding region
    - o Synonymous
    - o Nonsynonymous
        - Missense
        - Nonsense

Single-nucleotide polymorphisms may fall within coding sequences of genes, non-coding regions of genes, or in the intergenic regions (regions between genes). SNPs within a coding sequence do not necessarily change the amino acid sequence of the protein that is produced, due to degeneracy of the genetic code. A SNP in which both alleles produce the same polypeptide sequence is called a synonymous polymorphism (sometimes called a silent mutation). If a different polypeptide sequence is produced the polymorphism is a replacement polymorphism. A replacement polymorphism change may be either missense, which results in a different amino acid, or nonsense, which results in a premature stop codon. Over half of all known disease mutations come from replacement polymorphisms. SNPs that are not in protein-coding regions may still affect gene splicing, transcription factor binding, or the sequence of non-coding RNA. Gene expression affected by this type of SNP is referred to as an eSNP (expression SNP) and may be upstream or downstream from the gene.

Use and Importance

Variations in the DNA sequences of humans can affect how humans develop diseases and respond to pathogens, chemicals, drugs, vaccines, and other agents. SNPs are also thought to be key enablers in realizing the concept of personalized medicine. However, their greatest importance in biomedical research is for comparing regions of the genome between cohorts (such as with matched cohorts with and without a disease) in genome-wide association studies. They are usually biallelic and thus easily assayed. SNPs do not usually function individually, rather, they work in coordination with other SNPs to manifest a disease condition as has been seen in osteoporosis.

Single Nucleotide Variation Discovery

Transcriptome single nucleotide variation has been analyzed in maize on the Roche 454 sequencing platform. Directly from the transcriptome analysis, around 7000 single nucleotide polymorphisms (SNPs) were recognized. Following Sanger sequence validation, the researchers were able to conservatively obtain almost 5000 valid SNPs covering more than 2400 maize genes. This impressive transcriptome analysis is currently being applied to cancer research and microbiology which could quite possibly lead to new forms of medicine.

Coverage/Depth

Coverage/depth can affect the mutations seen and given that everything is expression-centric, an allele might not be detected either because it is not in the genome, or because it is not being expressed. At the same time, RNA-seq can yield additional information rather than just the existence of a heterozygous gene as it can also help in estimating the expression of each allele. In association studies, genotypes are associated to disease and expression levels can also be associated with disease. Using RNA-seq, we can measure the relationship between these two associated variables, that is, in what relation are each of the alleles being expressed. The depth of sequencing required for specific applications can be extrapolated from a pilot experiment.

Germline vs Expressed Alleles

The only way to be absolutely sure of the individual's mutations is to compare the transcriptome sequences to the germline DNA sequence. This enables the distinction of homozygous genes versus skewed expression of one of the alleles and it can also provide information about genes that were not expressed in the transcriptomic experiment.

Post-transcriptional SNVs

Having the matching genomic and transcriptomic sequences of an individual can also help in detecting post-transcriptional edits, where, if the individual is homozygous for a gene, but the gene's transcript has a different allele, then a post-transcriptional modification event is determined. mRNA centric single nucleotide variants (SNVs) are generally not considered as a representative source of functional variation in cells, mainly due to the fact that these mutations disappear with the mRNA molecule, however the fact that efficient DNA correction mechanisms do not apply to RNA molecules can cause them to appear more often. This has been proposed as the source of certain prion diseases, also known as TSE or transmissible spongiform encephalopathies.

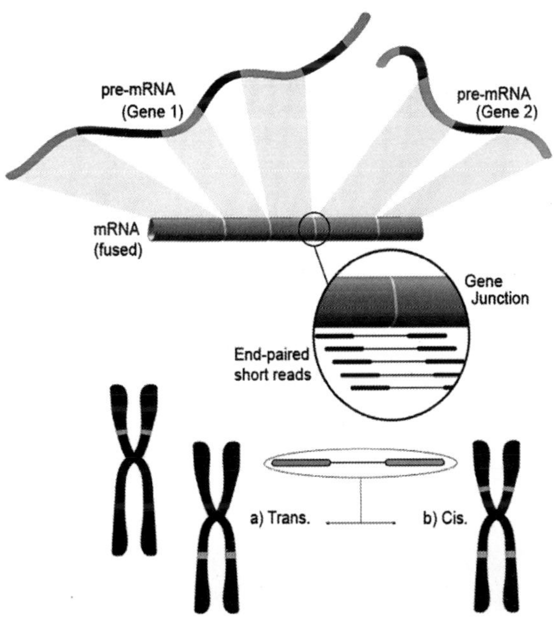

Figure: *RNA-seq mapping of short reads over exon-exon junctions, depending on where each end maps to, it could be defined a Trans or a Cis event.*

Fusion Gene

A fusion gene is a hybrid gene formed from two previously separate genes. It can occur as the result of a translocation, interstitial deletion, or chromosomal inversion. Often, fusion genes are oncogenes; examples include BCR-ABL, TEL-AML1 (ALL with t(12 ; 21)), AML1-ETO (M2 AML with t(8 ; 21)) and TMPRSS2-ERG with an interstitial deletion on chromosome 21, often occurring in prostate cancer. Most fusion genes are found from hematological cancers, sarcomas and prostate cancer.

Oncogenic fusion genes may lead to a gene product with a new or different function from the two fusion partners. Alternatively, a proto-oncogene is fused to a strong promoter, and thereby the oncogenic function is set to function by an upregulation caused by the strong promoter of the upstream fusion partner. The latter is common in lymphomas, where oncogenes are juxtaposed to the promoters of the immunoglobulin genes. Oncogenic fusion transcripts may also be caused by trans-splicing or read-through events .

Presence of certain chromosomal aberrations and their resulting fusion genes is commonly used within cancer diagnostics in order to set a precise diagnosis. In routine diagnostic laboratories the following methods are most commonly used: chromosome banding analysis, fluorescence in situ hybridisation, and RT-PCR. These methods all have their distinct shortcomings due to the very complex nature of cancer genomes. Recent developments such as high-throughput sequencing and custom-designed DNA microarrays bear promise of introduction of more efficient methods.

Biologists may also deliberately create fusion genes for research purposes. For example, by creating a fusion gene of a protein of interest and green fluorescent protein, the protein of interest may be observed in cells or tissue using fluorescence microscopy. The protein synthesized when a fusion gene is expressed is called a fusion protein.

Fusion Gene Detection

Caused by different structural modifications in the genome, fusion genes have gained attention because of their relationship with cancer. The ability of RNA-seq to analyze a sample's whole transcriptome in an unbiased fashion makes it an attractive tool to find these kinds of common events in cancer.

The idea follows from the process of aligning the short transcriptomic reads to a reference genome. Most of the short reads will fall within one complete exon, and a smaller but still large set would be expected to map to known exon-exon junctions. The remaining unmapped short reads would then be further analyzed to determine whether they match an exon-exon junction where the exons come from different genes.

This would be evidence of a possible fusion event, however, because of the length of the reads, this could prove to be very noisy. An alternative approach is to use pair-end reads, when a potentially large number of paired reads would map each end to a different exon, giving better coverage of these events.

Nonetheless, the end result consists of multiple and potentially novel combinations of genes providing an ideal starting point for further validation.

Some Considerations

The information gathered when sequencing a sample's transcriptome in this way has many of the same limitations as other RNA expression analysis pipelines. Mainly, the information gathered is:

a) Tissue specific: Gene expression is not uniform throughout an organism's cells, it is strongly dependent on the tissue type being measured;

b) Time dependent: During a cell's lifetime and context, its gene expression levels change.

Because of this, care must be taken when drawing conclusions from the sequencing experiment, as some of the information gathered might not be representative of the individual itself. An example of this would be during SNV discovery as the mutations discovered are more precisely the mutations being expressed, this is: observing a homozygote location to a non-reference allele in an organism does not necessarily mean that this is the individual's genotype, it could just mean that the gene copy with the reference allele is not being expressed in that tissue and/or at the time snapshot the sample was acquired.

Gene Family

A gene family is a set of several similar genes, formed by duplication of a single original gene, and generally with similar biochemical functions. One such family are the genes for human haemoglobin subunits; the ten genes are in two clusters on different chromosomes, called the á-globin and â-globin loci.

Genes are categorized into families based on shared nucleotide or protein sequences. Phylogenetic techniques can be used as a more rigorous test.

The positions of exons within the coding sequence can be used to infer common ancestry. Knowing the sequence of the protein encoded by a gene can allow researchers to apply methods that find similarities among protein sequences that provide more information than similarities or differences among DNA sequences. Furthermore, knowledge of the protein's secondary structure gives further information about ancestry, since the organization of secondary structural elements presumably would be conserved even if the amino acid sequence changes considerably. These methods often rely upon predictions based upon the DNA sequence.

If the genes of a gene family encode proteins, the term protein family is often used in an analogous manner to gene family.

The expansion or contraction of gene families along a specific lineage can be due to chance, or can be the result of natural selection.

To distinguish between these two cases is often difficult in practice. Recent work uses a combination of statistical models and algorithmic techniques to detect gene families that are under the effect of natural selection.

List of Gene Families

This is a list of gene families or gene complexes, that is sets of genes which occur across a number of different species which often serve similar biological functions. These gene families typically encode functionally related proteins, and sometimes the term gene families is a shorthand for the sets of proteins that the genes encode. They may or may not be physically adjacent on the same chromosome.

Genes-encoding Regulatory Proteins

- Homeobox (Hox gene family)
- Achaete-scute complex (Neuroblast formation)
- Krüppel-type zinc finger (ZNF)

Genes-encoding Immune System Proteins

- Major histocompatibility complex (MHC)

Related Protein Families

Motor Proteins

- Myosin
- Kinesin
- Dynein

Signal Transducing Proteins

- G-proteins
- MAP Kinase
- Olfactory Receptor
- Receptor tyrosine kinases

Transporters

- Cassidy proteins

To be sorted and annotated

Source

14-3-3 family

ABC Superfamily

ABC Transporters

ABC Transporters- PDR subfamily

AGC Kinase Gene Family

Aldehyde Dehydrogenase Gene Family

Amino Acid/Auxin Permease AAAP Family

Acyl Lipid Metabolism Gene Families

Antiporters

Aquaporins

ARIADNE family

AS2 family

basic Helix-Loop-Helix Transcription Factor Gene Family

Basic region leucine zipper (bZIP) Transcription Factor Gene Family

Calcium Dependent Protein Kinase

Calmodulin-binding Proteins

Carbohydrate Esterase

CBL-interacting serine-threonine protein kinases

Cell Wall Biosynthesis

Chloroplast and Mitochondria Gene Families

COBRA Gene Family

Core Cell Cycle Gene Families

Cytochrome P450

Cytochrome b5

Cytoplasmic Ribosomal Protein Gene Family

Cytoskeleton Gene Families

DOF Gene Family

EF-hand Containing Proteins

Equilibrative Nucleoside Transporter Gene Family (ENT)

Eukaryotic Initiation Factors Gene Family

Expansin Gene Family

F-Box Gene Family

FH2 protein (formin) Gene Family

FtsH Gene Family

Glycoside Hydrolase

Glycosyltransferase

GRAS Gene Family

Arabidopsis Glutathione S-transferase Gene Family

Heat shock transcription factors Hsfs

Histidine Kinase Gene Family

Histidine Phosphotransfer Proteins

IDZ Gene Family

Inorganic Solute Cotransporters

Ion Channels

Lateral Organ Boundaries Gene Family

Leucine-rich Repeat extensin Gene Families

Lipid Metabolism Gene Families

Kinesins gene family

Major Intrinsic Protein MIP

MAP Kinase Gene Family

MAP Kinase Kinase Gene Family

MAP Kinase Kinase Kinase

MAP Kinase Kinase Kinase Kinase

Miscellaneous Gene Families

MLO family

Monolignol Biosynthesis Gene Families

Monosaccharide Transporter(-like) Gene Family

MRS2 Gene Family

MYB family

Myosins gene family

NADPH P450 reductases

Nicotianamine Synthase Gene Family

Nodulin-like Gene Family

Nucleobase Ascorbate Transporter Gene Family

Organic Solute Cotransporters

Peroxidase, class III
Phospholipase D Gene Family (Zarsky group)
Phospholipase D Gene Family (Wang group)
Plant Defensins Gene Families
Polysaccharide Lyase
Pollen Coat Proteome
PP2C-type Phosphatase Gene Family
Primary Pumps ATPases
PRT Gene Family
RAD5/RAD16-like Gene Family
RCI2 Gene Family
Rapid Response to Elicitors
Receptor kinase-like Gene Family
Response Regulator Gene Family
SBP-box Gene Family
SNARE and Associated Proteins
Single Myb Histone (SMH) Gene Family
Subtilisin-like Serine Gene Family
Sucrose Transporter Gene Family
Sulfurtransferasese/Rhodanese
Strubbelig Receptor Gene Family
Trehalose Biosynthesis Gene Families
Type I MADS-box Gene Family
MADS like Gene Family
MADS-box Gene Family
U-box Gene Family
Wox Gene Family
WRKY Transcription Factor Superfamily
Xyloglucan Fucosyltransferase1
Zinc finger-homeobox gene family

Gene Patent

A gene patent is a patent on a specific isolated gene sequence, its chemical composition, the processes for obtaining or using it, or a combination of such claims. Gene patents are a part of the broader category of biological patents.

Gene patents may claim the isolated natural sequences of genes, the use of a natural sequence for purposes such as diagnostic testing, or a natural sequence that has been altered by adding a promoter or other changes to make it more useful. The courts have upheld claims on altered sequences, but courts and lawyers' opinions have been mixed on upholding the use of natural sequences and particularly the sequence itself. Patents on genes have only been granted on isolated gene sequences with known functions, and these patents cannot be applied to the naturally occurring genes in humans or any other naturally occurring organism.

History

The United States has been patenting chemical compositions based upon human products for over 100 years. The first patent for a human product was granted on March 20, 1906 for a purified form of adrenaline. It was challenged and upheld in Parke-Davis v. Mulford Judge Hand argued that natural substances when they are purified are more useful than the original natural substances.

In 1980, the U.S. Supreme Court, in Diamond v. Chakrabarty, upheld the first patent on a newly-created living organism, a bacterium for digesting crude oil in oil spills. The patent examiner for the United States Patent and Trademark Office had rejected the patent of a living organism, but Chakrabarty appealed. As a rule, raw natural material is generally rejected for patent approval by the USPTO. The Court ruled that as long as the organism is truly "man-made," such as through genetic engineering, then it is patentable. Because the DNA of Chakrabarty's organism was modified, it was patentable.

Controversy

Law professor Rebecca S. Eisenberg argues that gene patents produce an "anticommons" at odds with an ideal scientific commons. Others claim that patents have not created this "anticommons" effect on research, based on surveys of scientists. Professional societies of pathologists have criticized patents on disease genes and exclusive licenses to perform DNA diagnostic tests. In the 2009 Myriad case, doctors and pathologists complained that the patent on BRCA1 and BRCA2 genes prevented patients from receiving second opinions on their test results. Pathologists complained that the patent prevented them from carrying out their medical practice of doing diagnostic tests on patient samples and interpreting the results.

Another example is a series of lawsuits filed by the Alzheimer's Institute of America (AIA) starting in 2010, concerning a gene patent it controls. The patent covers a genetic mutation that predisposes to Alzheimer's, and more importantly, it applies to transgenic mice carrying the mutation. These mice are widely used in Alzheimer's research, both by academic scientists doing basic research and by companies that use the mice to test products in development. Two of these suits are directed to companies that were started based on inventions made at universities (Comentis and Avid), and in each of those cases, the university has been sued along with the company. While none of the suits target universities that are conducting basic research using the mice, one of the suits is against Jackson Labs, a nonprofit company that provides transgenic mice to academic and commercial researchers and is an important repository of such mice.

While there is some controversy concerning the patenting of isolated genes and the way those patents are used, and there is controversy concerning patents on the diagnostic uses of genes (the real source of dispute in the Myriad court case), it is difficult to find controversy surrounding patents on genes that are used to manufacture therapeutic proteins (for an example of patents on therapeutic proteins, the drug candidate that is the subject of the early part of the movie 'Extraordinary Measures' was covered in part by a classic gene patent, US Patent 6,770,468.) There is also little controversy concerning the role of gene patents in the chemical industry—for instance in the manufacture of enzymes used in consumer products or industrial processes. In descriptions of the controversy over "gene patents", it is rare to find references to these everyday uses of genes and gene patents.

4

Synthesis by the Phosphoramidite Method

Building Blocks

Nucleoside phosphoramidites

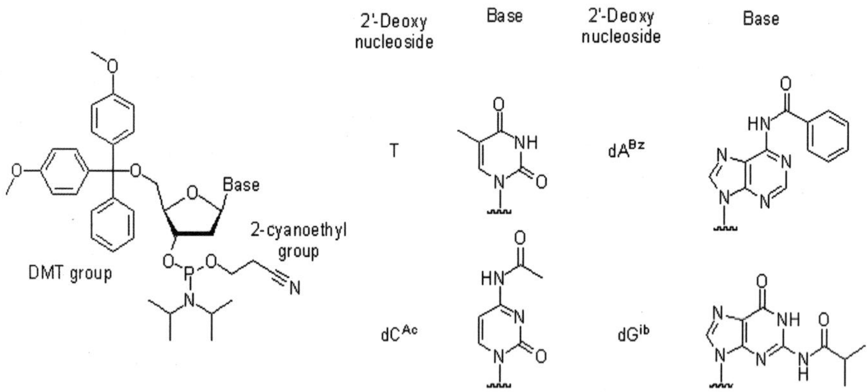

***Figure:** Protected 2'-deoxynucleoside phosphoramidites.*

As mentioned above, the naturally occurring nucleotides (nucleoside-3'- or 5'-phosphates) and their phosphodiester analogs are insufficiently reactive to afford an expedite synthetic preparation of oligonucleotides in high yields. The selectivity and the rate of the formation of internucleosidic linkages is dramatically improved by using 3'-O-(N,N-diisopropyl phosphoramidite) derivatives of nucleosides (nucleoside phosphoramidites) that serve as building blocks in phosphite triester methodology. To prevent undesired side reactions, all other functional groups present in nucleosides have to be rendered unreactive (protected) by attaching protecting groups. Upon the completion of the oligonucleotide chain assembly, all the protecting groups are removed to yield the desired oligonucleotides. Below, the protecting groups currently used in commercially available and most common nucleoside phosphoramidite building blocks are briefly reviewed:

- The 5'-hydroxyl group is protected by an acid-labile DMT (4,4'-dimethoxytrityl) group.
- Thymine and uracil, nucleic bases of thymidine and uridine, respectively, do not have exocyclic amino groups and hence do not require any protection. Although the nucleic base of guanosine and 2'-deoxyguanosine does have an exocyclic amino group, its basicity is low to an extent that it does not react with phosphoramidites under the conditions of the coupling reaction. However, a phosphoramidite derived from the N2-unprotected 5'-O-DMT-2'-deoxyguanosine is poorly soluble in acetonitrile, the solvent commonly used in oligonucleotide synthesis. In contrast, the N2-protected versions of the same compound dissolve in acetonitrile well and hence are widely used. Nucleic bases adenine and cytosine bear the exocyclic amino groups reactive with the activated phosphoramidites under the conditions of the coupling reaction. Although, at the expense of additional steps in the synthetic cycle, the oligonucleotide chain assembly may be carried out using dA and dC phosphoramidites with unprotected amino groups, most often these are kept permanently protected over the entire length of the oligonucleotide chain assembly. The protection of the exocyclic amino groups have to be orthogonal to that of the 5'-hydroxy group because the latter is removed at the end of each synthetic cycle. The simplest to implement and hence the most widely accepted is the strategy where the exocyclic amino groups bear a base-labile protection. Most often, two protection schemes are used.
- In the first, the standard and more robust scheme, Bz (benzoyl) protection is used for A, dA, C, and dC, while G and dG are protected with isobutyryl group. More recently, Ac (acetyl) group is often used to protect C and dC.
- In the second, mild protection scheme, A and dA are protected with isobutyryl or phenoxyacetyl groups (PAC). C and dC bear acetyl protection, and G and dG are protected with 4-isopropylphenoxyacetyl (iPr-PAC) or dimethylformamidino (dmf) groups. Mild protecting groups are removed more readily than the standard protecting groups. However, the phosphoramidites bearing these groups are less stable when stored in solution.

- The phosphite group is protected by a base-labile 2-cyanoethyl group. Once a phosphoramidite has been coupled to the solid support-bound oligonucleotide and the phosphite moieties have been converted to the P(V) species, the presence of the phosphate protection is not mandatory for the successful conducting of further coupling reactions.

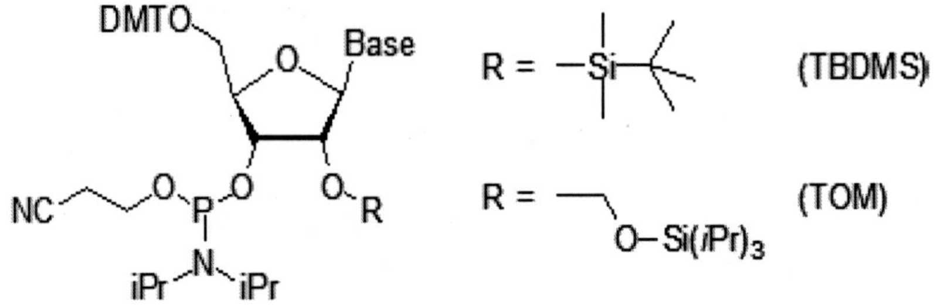

Figure: *2'-O-Protected ribonucleoside phosphoramidites.*

- In RNA synthesis, the 2'-hydroxy group is protected with TBDMS (t-butyldimethylsilyl) group. or with TOM (tri-iso-propylsilyloxymethyl) group, both being removable by treatment with fluoride ion.
- The phosphite moiety also bears a diisopropylamino (iPr_2N) group reactive under acidic conditions. Upon activation, the diisopropylamino group leaves to be substituted by the 5'-hydroxy group of the support-bound oligonucleotide.

Synthetic Cycle

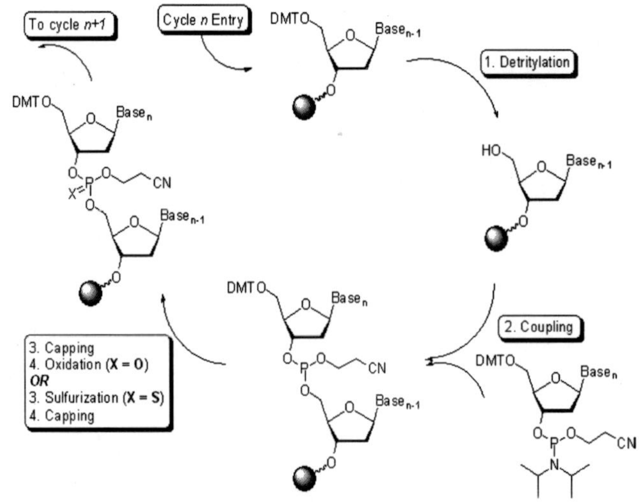

Figure: *Synthetic cycle for preparation of oligonucleotides by phosphoramidite method.*

Oligonucleotide synthesis is carried out by a stepwise addition of nucleotide residues to the 5'-terminus of the growing chain until the desired sequence is assembled. Each addition is referred to as a synthetic cycle (Scheme 6) and consists of four chemical reactions:

Step 1: De-blocking (detritylation): The DMT group is removed with a solution of an acid, such as 2% TCA or 3% Dichloroacetic acid (DCA), in an inert solvent (dichloromethane or toluene). The orange-colored DMT cation formed is washed out; the step results in the solid support-bound oligonucleotide precursor bearing a free 5'-terminal hydroxyl group. It is worth remembering that conducting detritylation for an extended time or with stronger than recommended solutions of acids leads to depurination of solid support-bound oligonucleotide and thus reduces the yield of the desired full-length product.

Step 2: Coupling: A 0.02 - 0.2M solution of nucleoside phosphoramidite (or a mixture of several phosphoramidites) in acetonitrile is activated by a 0.2 - 0.7M solution of an acidic azole catalyst, 1H-tetrazole, 2-ethylthiotetrazole, 2-benzylthiotetrazole, 4,5-dicyanoimidazole, or a number of similar compounds. The mixing is usually very brief and occurs in fluid lines of oligonucleotide synthesizers while the components are being delivered to the reactors containing solid support.

The activated phosphoramidite in 1.5 - 20-fold excess over the support-bound material is then brought in contact with the starting solid support (first coupling) or a support-bound oligonucleotide precursor (following couplings) whose 5'-hydroxy group reacts with the activated phosphoramidite moiety of the incoming nucleoside phosphoramidite to form a phosphite triester linkage. The phosphoramidite coupling is very rapid and requires, on small scale, about 20 s for its completion. The reaction is also highly sensitive to the presence of water, particularly when dilute solutions of phosphoramidites are used, and is commonly carried out in anhydrous acetonitrile. Generally, the larger the scale of the synthesis, the lower the excess and the higher the concentration of the phosphoramidites is used. In contrast, the concentration of the activator is primarily determined by its solubility in acetonitrile and is irrespective of the scale of the synthesis. Upon the completion of the coupling, any unbound reagents and by-products are removed by washing.

Step 3: Capping: The capping step is performed by treating the solid support-bound material with a mixture of acetic anhydride and 1-methylimidazole or, less often, DMAP as catalysts and, in the phosphoramidite method, serves two purposes.

- After the completion of the coupling reaction, a small percentage of the solid support-bound 5'-OH groups (0.1 to 1%) remains unreacted and needs to be permanently blocked from further chain elongation to prevent the formation of oligonucleotides with an internal base deletion commonly referred to as (n-1) shortmers. The unreacted 5'-hydroxy groups are, to a large extent, acetylated by the capping mixture.

- It has also been reported that phosphoramidites activated with 1H-tetrazole react, to a small extent, with the O position of guanosine. Upon oxidation with I_2 /water, this side product, possibly via O -N7 migration, undergoes depurination. The apurinic sites thus formed are readily cleaved in the course of the final deprotection of the oligonucleotide under the basic conditions to give two shorter oligonucleotides thus reducing the yield of the full-length product. The O modifications are rapidly removed by treatment with the capping reagent as long as the capping step is performed prior to oxidation with I_2/water.
- The synthesis of oligonucleotide phosphorothioates does not involve the oxidation with I_2/water, and, respectively, does not suffer from the side reaction described above. On the other hand, if the capping step is performed prior to sulfurization, the solid support may contain the residual acetic anhydride and N-methylimidazole left after the capping step.

  The capping mixture interferes with the sulfur transfer reaction, which results in the extensive formation of the phosphate triester internucleosidic linkages in place of the desired PS triesters. Therefore, for the synthesis of OPS, it is advisable to conduct the sulfurization step prior to the capping step.

Step 4: Oxidation: The newly formed tricoordinated phosphite triester linkage is not natural and is of limited stability under the conditions of oligonucleotide synthesis.

The treatment of the support-bound material with iodine and water in the presence of a weak base (pyridine, lutidine, or collidine) oxidizes the phosphite triester into a tetracoordinated phosphate triester, a protected precursor of the naturally occurring phosphate diester internucleosidic linkage. This step is substituted with a sulfurization step to obtain oligonucleotide phosphorothioates. In the latter case, the sulfurization step is best carried out prior to capping.

Solid Supports

In solid-phase synthesis, an oligonucleotide being assembled is covalently bound, via its 3'-terminal hydroxy group, to a solid support material and remains attached to it over the entire course of the chain assembly. The solid support is contained in columns whose dimensions depend on the scale of synthesis and may vary between 0.05 mL and several litres.

The overwhelming majority of oligonucleotides are synthesized on small scale ranging from 40 nmol to 1 ìmol. More recently, high-throughput oligonucleotide synthesis where the solid support is contained in the wells of multi-well plates (most often, 96 or 384 wells per plate) became a method of choice for parallel synthesis of oligonucleotides on small scale. At the end of the chain assembly, the oligonucleotide is released from the solid support and eluted from the column or the well.

Solid Support Material

In contrast to organic solid-phase synthesis and peptide synthesis, the synthesis of oligonucleotides proceeds best on non-swellable or low-swellable solid supports. The two most often used solid-phase materials are Controlled Pore Glass (CPG) and macroporous polystyrene (MPPS).

- CPG is commonly defined by its pore size. In oligonucleotide chemistry, pore sizes of 500, 1000, 1500, 2000, and 3000 Å are used to allow the preparation of about 50, 80, 100, 150, and 200-mer oligonucleotides, respectively. To make native CPG suitable for further processing, the surface of the material is treated with (3-aminopropyl)triethoxysilane to give Aminopropyl CPG. The aminopropyl arm may be further extended to result in Long Chain Aminoalkyl (LCAA) CPG. The amino group is then used as an anchoring point for linkers suitable for oligonucleotide synthesis.
- MPPS suitable for oligonucleotide synthesis is a low-swellable, highly cross-linked polystyrene obtained by polymerization of divinylbenzene (min 60%), styrene, and 4-chloromethylstyrene in the presence of a porogeneous agent. The macroporous chloromethyl MPPS obtained is converted to aminomethyl MPPS.

Linker Chemistry

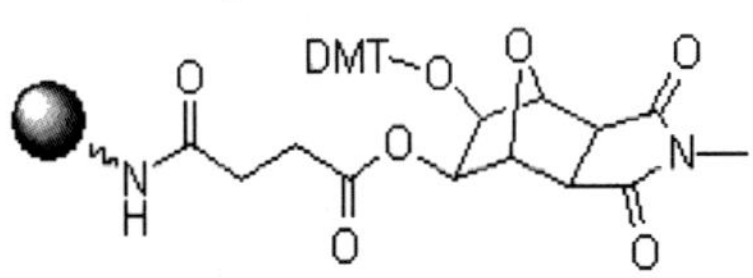

1

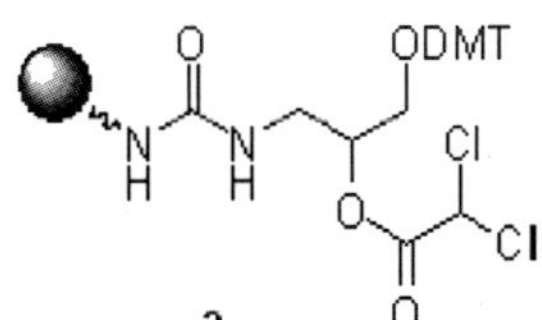

2

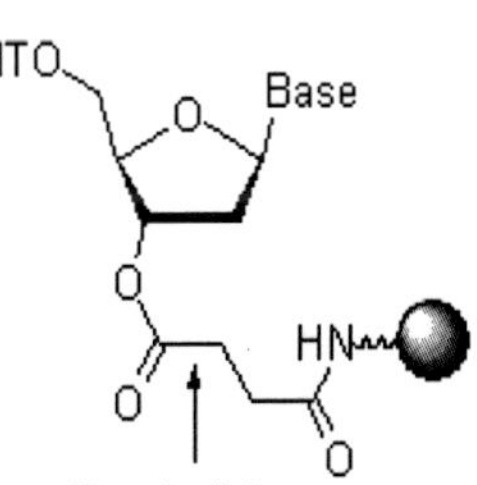

3

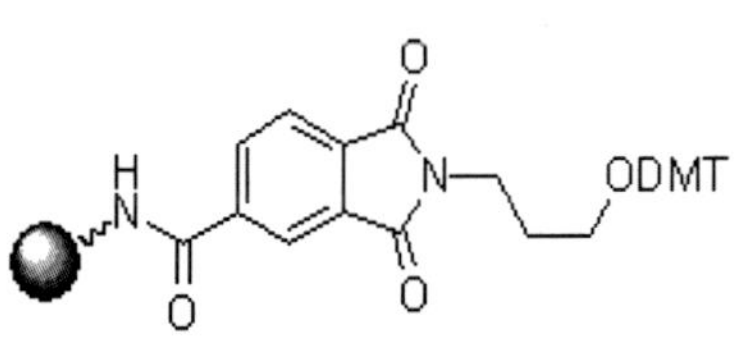

4

Figure: *Commercial solid supports for oligonucleotide synthesis.*

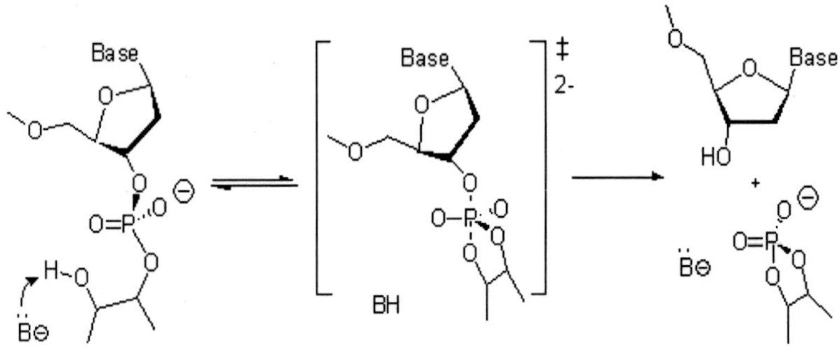

Figure: *Mechanism of 3'-dephosphorylation of oligonucleotides assembled on universal solid supports.*

To make the solid support material suitable for oligonucleotide synthesis, non-nucleosidic linkers or nucleoside succinates are covalently attached to the reactive amino groups in Aminopropyl CPG, LCAA CPG, or Aminomethyl MPPS.

The remaining unreacted amino groups are capped with acetic anhydride. Typically, three conceptually different groups of solid supports are used.

- Universal supports. In a more recent and more convenient method, the synthesis starts with the universal support where a non-nucleosidic linker is attached to the solid support material (Compounds 1 and 2). A phosphoramidite respective to the 3'-terminal nucleoside residue is coupled to the universal solid support in the first synthetic cycle of oligonucleotide chain assembly using the standard protocols. The chain assembly is then continued until the completion, after which the solid support-bound oligonucleotide is deprotected. The characteristic feature of the universal solid supports is that the release of the oligonucleotides occurs by the hydrolytic cleavage of a P-O bond that attaches the 3'-O of the 3'-terminal nucleotide residue to the universal linker as shown in Scheme 5. The critical advantage of this approach is that the same solid support is used irrespectively of the sequence of the oligonucleotide to be synthesized. For the complete removal of the linker and the 3'-terminal phosphate from the assembled oligonucleotide, the solid support 1 and several similar solid supports require gaseous ammonia, aqueous ammonium hydroxide, aqueous methylamine, or their mixture and are commercially available. The solid support 2 requires a solution of ammonia in anhydrous methanol and is also commercially available.
- Nucleosidic solid supports. In a historically first and still popular approach, the 3'-hydroxy group of the 3'-terminal nucleoside residue is attached to the solid support via, most often, 3'-O-succinyl arm as in compound 3. The oligonucleotide chain assembly starts with the coupling of a phosphoramidite building block respective to the

nucleotide residue second from the 3'-terminus. The 3'-terminal hydroxy group in oligonucleotides synthesized on nucleosidic solid supports is deprotected under the conditions somewhat milder than those applicable for universal solid supports. However, the fact that a nucleosidic solid support has to be selected in a sequence-specific manner reduces the throughput of the entire synthetic process and increases the likelihood of human error.

- Special solid supports are used for the attachment of desired functional or reporter groups at the 3'-terminus of synthetic oligonucleotides. For example, the commercial solid support 4 allows the preparation of oligonucleotides bearing 3'-terminal 3-aminopropyl linker. Similarly to non-nucleosidic phosphoramidites, many other special solid supports designed for the attachment of reactive functional groups, non-radioactive reporter groups, and terminal modifiers (e.c., hydrophobic tethers) and suited for various applications are commercially available.

Oligonucleotide Phosphorothioates and Their Synthesis

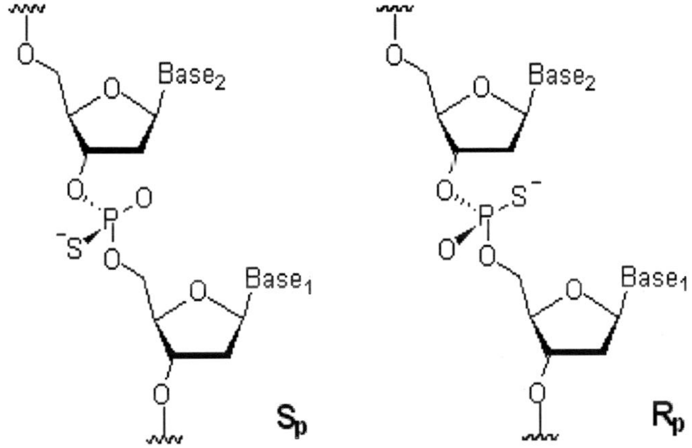

Figure: *S_p and R_p-diastereomeric internucleosidic phosphorothioate linkages.*

Oligonucleotide phosphorothioates (OPS) are modified oligonucleotides where one of the oxygen atoms in the phosphate moiety is replaced by sulfur. Only the phosphorothioates having sulfur at a non-bridging position as shown in figure are widely used and are available commercially. The replacement of the non-bridging oxygen with sulfur creates a new center of chirality at phosphorus. In a simple case of a dinucleotide, this results in the formation of a diastereomeric pair of S_p- and R_p-dinucleoside monophosphorothioates whose structures. In a n-mer oligonucleotide where all (n – 1) internucleosidic linkages are phosphorothioate linkages, the number of diastereomers m is calculated as m = 2 . Being non-natural analogs of nucleic acids, OPS are substantially more stable towards hydrolysis by nucleases, the class of enzymes that destoy nucleic acids by breaking the bridging P-O bond of the phosphodiester moiety. This property determines the use of OPS as antisense oligonucleotides in in vitro and in vivo applications where the extensive exposure to nucleases is inevitable. Similarly, to improve the stability of siRNA, at least one phosphorothioate linkage is often introduced at the 3'-terminus of both sense and antisense strands. In chirally pure OPS, all-Sp diastereomers are more stable to enzymatic degradation than their all-Rp analogs. However, the preparation of chirally pure OPS remains a synthetic challenge. In laboratory practice, mixtures of diastereomers of OPS are commonly used. Synthesis of OPS is very similar to that of natural oligonucleotides. The difference is that the oxidation step is replaced by sulfur transfer reaction (sulfurization) and that the capping step is performed after the sulfurization. Of many reported reagents capable of the efficient sulfur transfer, only three are commercially available:

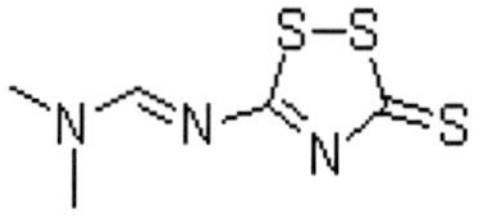

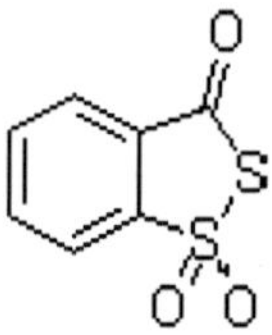

3 **4**

Figure: *Commercial sulfur transfer agents for oligonucleotide synthesis.*

- 3-((Dimethylaminomethylidene)amino)-3H-1,2,4-dithiazole-3-thione, DDTT (3, Fig. 4) provides rapid kinetics of sulfurization and high stability in solution. The reagent is available from several sources.
- 3H-1,2-benzodithiol-3-one 1,1-dioxide (4, Fig. 4) also known as Beaucage Reagent displays a better solubility in acetonitrile and short reaction times. However, the reagent is of limited stability in solution and is less efficient in sulfurizing RNA linkages.
- N,N,N'N'-Tetraethylthiuram disulfide (TETD) is soluble in acetonitrile and is commercially available. However, the sulfurization reaction of an internucleosidic DNA linkage with TETD requires 15 min, which is more than 10 times as slow as that with compounds 3 and 4.

Automation

In the past, oligonucleotide synthesis was carried out manually in solution or on solid phase. The solid phase synthesis was implemented using, as containers for the solid phase, miniature glass columns similar in their shape to low-pressure chromatography columns or syringes equipped with porous filters. Currently, solid-phase oligonucleotide synthesis is carried out automatically using computer-controlled instruments (oligonucleotide synthesizers) and is technically implemented in column, multi-well plate, and array formats.

The column format is best suited for research and large scale applications where a high-throughput is not required. Multi-well plate format is designed specifically for high-throughput synthesis on small scale to satisfy the growing demand of industry and academia for synthetic oligonucleotides. A number of oligonucleotide synthesizers for small scale synthesis and medium to large scale synthesis are available commercially.

Synthesis of Oligonucleotide Microarrays

One may visualize an oligonucleotide microarray as a miniature multi-well plate where physical dividers between the wells (plastic walls) are intentionally removed. With respect to the chemistry, synthesis of oligonucleotide microarrays is different from the conventional oligonucleotide synthesis in two respects:

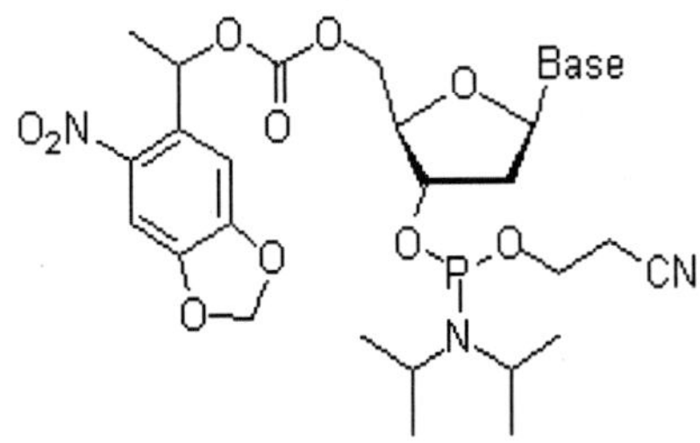

Figure: *5'-O-NPPOC-protected nucleoside phosphoramidite.*

- Oligonucleotides remain permanently attached to the solid phase, which requires the use of linkers that are stable under the conditions of the final deprotection procedure.
- The absence of physical dividers between the sites occupied by individual oligonucleotides, a very limited space on the surface of the microarray (one oligonucleotide sequence occupies a square 25x25 ìm) and the requirement of high fidelity of oligonucleotide synthesis dictate the use of site-selective 5'-deprotection techniques. In one approach, the removal of the 5'-O-DMT group is effected by electrochemical generation of the acid at the required site(s). In another approach, 5'-O-(á-methyl-6-nitropiperonyloxycarbonyl) (NPPOC) protecting group removable by UV-irradiation at 365 nm is used.

Post-synthetic processing

After the completion of the chain assembly, the solid support-bound oligonucleotide is fully protected:

- The 5'-terminal 5'-hydroxy group is protected with DMT group;
- The internucleosidic phosphate or phosphorothioate moieties are protected with 2-cyanoethyl groups;
- The exocyclic amino groups in all nucleis bases except for T and U are protected with acyl protecting groups.

To furnish a functional oligonucleotide, all the protecting groups have to be removed. The N-acyl base protection and the 2-cyanoethyl phosphate protection may be, and is often removed simultaneously by treatment with inorganic bases or amines. However, the applicability of this method is limited by the fact that the cleavage of 2-cyanoethyl phosphate protection gives rise to acrylonitrile as a side product. Under the strong basic conditions required for the removal of N-acyl protection, acrylonitrile is capable of alkylation of nucleic bases, primarily, at the N3-position of thymine and uracil residues to give the respective N3-(2-cyanoethyl) adducts via Michael reaction. The formation of these side products may be avoided by treating the solid support-boud oligonucleotides with solutions of bases in an organic solvent, for instance, with 50% triethylamine in acetonitrile or 10% diethylamine in acetonitrile.

This treatment is strongly recommended for medium- and large scale preparations and is optional for syntheses on small scale where the concentration of acrylonitrile generated in the deprotection mixture is low. Regardless of whether the phosphate protecting groups were removed first, the solid support-bound oligonucleotides are deprotected using one of the two general approaches.

- (1) Most often, 5'-DMT group is removed at the end of the oligonucleotide chain assembly. The oligonucleotides are then released from the solid phase and deprotected (base and phosphate) by treatment with aqueous ammonium hydroxide, aqueous methylamine, their mixtures, gaseous ammonia or methylamine or, less commonly, solutions of other primary amines or alkalies at ambient or elevated temperature. This removes all remaining protection groups from 2'-deoxyoligonucleotides, resulting in a reaction mixture containing the desired product. If the oligonucleotide contains any 2'-O-protected ribonucleotide residues, the deprotection protocol includes the second step where the 2'-O-protecting silyl groups are removed by treatment with fluoride ion by various methods.

The fully deprotected product is used as is, or the desired oligonucleotide can be purified by a number of methods. Most commonly, the crude product is desalted using ethanol precipitation, size exclusion chromatography, or reverse-phase HPLC. To eliminate unwanted truncation products, the oligonucleotides can be purified via polyacrylamide gel electrophoresis or anion-exchange HPLC followed by desalting.

- (2) The second approach is only used when the intended method of purification is reverse-phase HPLC. In this case, the 5'-terminal DMT group that serves as a hydrophobic handle for purification is kept on at the end of the synthesis. The oligonucleotide is deprotected under basic conditions as described above and, upon evaporation, is purified by reverse-phase HPLC. The collected material is then detritylated under aqueous acidic conditions. On small scale (less than 0.01–

0.02 mmol), the treatment with 80% aqueous acetic acid for 15–30 min at room temperature is often used followed by evaporation of the reaction mixture to dryness in vacuo. Finally, the product is desalted as described above.

- For some applications, additional reporter groups may be attached to an oligonucleotide using a variety of post-synthetic procedures.

Characterization

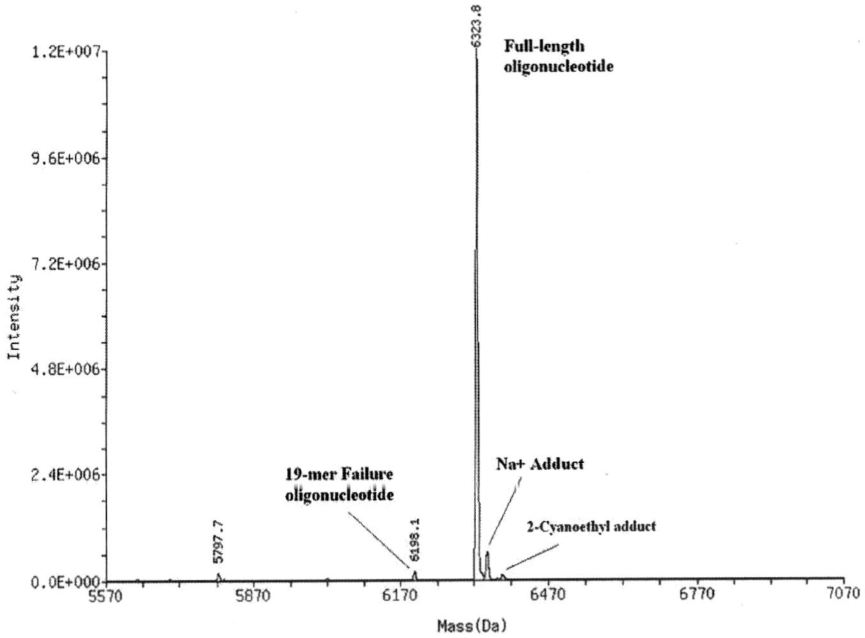

Figure: *Deconvoluted ES MS of crude oligonucleotide 5'-DMT-T_{20} (Calcd. 6324.26 Da).*

As with any other organic compound, it is prudent to characterize synthetic oligonucleotides upon their preparation. In more complex cases (research and large scale syntheses) oligonucleotides are characterized after their deprotection and after purification. Although the ultimate approach to the characterization is sequencing, a relatively inexpensive and routine procedure, the considerations of the cost reduction preclude its use in routine manufacturing of oligonucleotides. In day-by-day practice, it is sufficient to obtain the molecular mass of an oligonucleotide by recording its mass spectrum. Two methods are currently widely used for characterization of oligonucleotides: electrospray mass spectrometry (ES MS) and matrix-assisted laser desorption/ionization time-of-flight mass spectrometry (MALDI-TOF). Prior to submitting a sample to the analysis by either of the methods, it is very important to exchange all metal ions that might be present in the sample for ammonium or trialkylammonium (e.c. triethylammonium) ions.

- In ES MS spectrum, a given oligonucleotide generates a set of ions that correspond to different ionization states of the compound. Thus, the oligonucleotide with molecular mass M generates ions with masses (M – nH)/n where M is the molecular mass of the oligonucleotide in the form of a free acid (all negative charges of internucleosidic phosphodiester groups are neutralized with H), n is the ionization state, and H is the atomic mass of hydrogen (1 Da). Most useful for characterization are the ions with n ranging from 2 to 5. Software supplied with the more recently manufactured instruments is capable of performing a deconvolution procedure that is, it finds peaks of ions that belong to the same set and derives the molecular mass of the oligonucleotide.
- To obtain more detailed information on the impurity profile of oligonucleotides, liquid chromatography-mass spectrometry (LC-MS or HPLC-MS) or capillary electrophoresis mass spectrometry (CEMS) are used.
- Antisense

In molecular biology and genetics, sense is a concept used to compare the polarity of nucleic acid molecules, such as DNA or RNA, to other nucleic acid molecules. Depending on the context within molecular biology, sense may have slightly different meanings.

DNA Sense

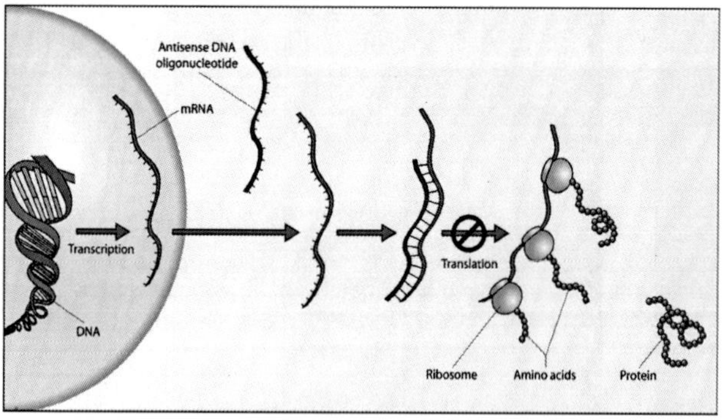

***Figure:** Schematic showing how antisense DNA strands can interfere with protein translation.*

Molecular biologists call a single strand of DNA sense (or positive (+) sense) if an RNA version of the same sequence is translated or translatable into protein. Its complementary strand is called antisense (or negative (-) sense). Sometimes the phrase coding strand is encountered; however, protein coding and non-coding RNA's can be transcribed similarly from both strands, in some cases being transcribed in both directions from a common promoter region, or being transcribed from within introns, on both strands.

Antisense DNA

The two complementary strands of double-stranded DNA (dsDNA) are usually differentiated as the "sense" strand and the "antisense" strand. The DNA sense strand looks like the messenger RNA (mRNA) and can be used to read the expected protein code by human eyes (e.g. ATG codon = Methionine amino acid). However, the DNA sense strand itself is not used to make protein by the cell. It is the DNA antisense strand which serves as the source for the protein code, because, with bases complementary to the DNA sense strand, it is used as a template for the mRNA. Since transcription results in an RNA product complementary to the DNA template strand, the mRNA is complementary to the DNA antisense strand. The mRNA is what is used for translation (protein synthesis).

Hence, a base triplet 3'-TAC-5' in the DNA antisense strand can be used as a template which will result in an 5'-AUG-3' base triplet in mRNA (AUG is the codon for Methionine, the start codon). The DNA sense strand will have the triplet ATG which looks just like AUG but will not be used to make Methionine because it will not be used to make mRNA. The DNA sense strand is called a "sense" strand not because it will be used to make protein (it won't be), but because it has a sequence that looks like the protein codon sequence.

Students may find this confusing if they misunderstand the meaning of "sense" and if they misunderstand complementation. To make things more confusing, before the convention was set, very early textbooks disagreed on the DNA strands called "sense" and "antisense."

In biology and research, short antisense molecules can interact with complementary strands of nucleic acids, modifying expression of genes.

Example with Double-stranded DNA

DNA strand 1: antisense strand (copied to)'! RNA strand (sense)

DNA strand 2: sense strand

Some regions within a double strand of DNA code for genes, which are usually instructions specifying the order of amino acids in a protein along with regulatory sequences, splicing sites, noncoding introns, and other complicating details. For a cell to use this information, one strand of the DNA serves as a template for the synthesis of a complementary strand of RNA. The template DNA strand is called the transcribed strand with antisense sequence and the mRNA transcript is said to be sense sequence (the complement of antisense). Because the DNA is double-stranded, the strand complementary to the antisense sequence is called non-transcribed strand and has the same sense sequence as the mRNA transcript (though T bases in DNA are substituted with U bases in RNA).

Table 1:

5'CGCTATAG CGTTTCAT 3'	DNA antisense strand (template/noncoding), Watson strand	Used as a template for transcription.
3'GCGATAT CGCAAAGTA 5'	DNA sense strand (nontemplate/coding), Crick strand	Complementary to the template strand.
3'GCGAUAUCG CAAAGUA 5'	mRNA Sense transcript	RNA strand that is transcribed from the noncoding (template/antisense) strand. Note : Except for the fact that all thymines are now uracils (T—>U), it is complementary to the noncoding (template/ antisense) DNA strand (identical to the coding (nontemplate/sense) DNA strand). Note There is an AUG start codon at the 5' end (although written backwards here).
5'CGCUAUAGC GUUUCAU 3'		mRNA Antisense transcript RNA strand that is transcribed from the coding (nontemplate/sense) strand. Note: Except for the fact that all thymines are now uracils (T—>U), it is complementary to the coding (nontemplate/sense) DNA strand (identical to the noncoding (template/ antisense) DNA strand.

A note on the confusion between "sense" and "antisense" strands: The strand names actually depend on which direction you are writing the sequence that contains the information for proteins (the "sense" information), not on which strand is on the top or bottom (that is arbitrary). The only real biological information that is important for labeling strands is the location of the 5' phosphate group and the 3' hydroxyl group because these ends determine the direction of transcription and translation. A sequence 5' CGCTAT 3' is equivalent to a sequence written 3' TATCGC 5' as long as the 5' and 3' ends are noted. If the ends are not labeled, convention is to assume that the sequence is written in the 5' to 3' direction. Good rule of thumb for figuring out the "sense" strand: Look for the start codon ATG (AUG in mRNA). In the table example, the sense mRNA has the AUG codon at the end (remember that translation proceeds in the 5' to 3' direction).

Ambisense

A single-stranded genome that contains both positive-sense and negative-sense is said to be ambisense. Bunya viruses have 3 single-stranded RNA (ssRNA) fragments containing both positive-sense and negative-sense sections; arenaviruses are also ssRNA viruses with an ambisense genome, as they have 2 fragments that are mainly negative-sense except for part of the 5' ends of the large and small segments of their genome.

Antisense RNA

Antisense RNA is an RNA transcript that is complementary to endogenous mRNA. In other words, it is a non-coding strand complementary to the coding sequence of RNA; this is similar to negative-sense viral RNA. Introducing a transgene coding for antisense RNA is a technique used to block expression of a gene of interest. Radioactively-labelled antisense RNA can be used to show the level of transcription of genes in various cell types. Some alternative antisense structural types are being experimentally applied as antisense therapy, with at least one antisense therapy approved for use in humans.

When mRNA forms a duplex with a complementary antisense RNA sequence, translation is blocked. This process is called RNA interference. Antisense nucleic acid molecules have been used experimentally to bind to mRNA and prevent expression of specific genes. Antisense therapies are also in development; in the USA, the Food and Drug Administration (FDA) has approved a phosphorothioate antisense oligo, fomivirsen (Vitravene), for human therapeutic use. Cells can produce antisense RNA molecules naturally, which interact with complementary mRNA molecules and inhibit their expression.

RNA Sense in Viruses

In virology, the genome of an RNA virus can be said to be either positive-sense, also known as a "plus-strand", or negative-sense, also known as a "minus strand". In most cases, the terms sense and strand are used interchangeably, making such terms as positive-strand equivalent to positive-sense, and plus-strand equivalent to plus-sense. Whether a virus genome is positive-sense or negative-sense can be used as a basis for classifying viruses.

Positive-sense

Positive-sense (5' to 3') viral RNA signifies that a particular viral RNA sequence may be directly translated into the desired viral proteins. Therefore, in positive-sense RNA viruses, the viral RNA genome can be considered viral mRNA, and can be immediately translated by the host cell. Unlike negative-sense RNA, positive-sense RNA is of the same sense as mRNA.

Some viruses (e.g., Coronaviridae) have positive-sense genomes that can act as mRNA and be used directly to synthesize proteins without the help of a complementary RNA intermediate. Because of this, these viruses do not need to have an RNA polymerase packaged into the virion.

Negative-sense

Negative-sense (3' to 5') viral RNA is complementary to the viral mRNA and thus must be converted to positive-sense RNA by an RNA polymerase prior to translation. Negative-sense RNA (like DNA) has a nucleotide sequence complementary to the mRNA that it encodes. Like DNA, this RNA cannot be translated into protein directly. Instead, it must first be transcribed into a positive-sense RNA that acts as an mRNA. Some viruses (Influenza, for example) have negative-sense genomes and so must carry an RNA polymerase inside the virion.

Enzymes

Enzyme-dependent antisense includes forms dependent on RNase H activity to degrade target mRNA, including single-stranded DNA, RNA, and phosphorothioate antisense. The R1 plasmid hok/sok system is an example of mRNA antisense regulation process, through enzymatic degradation of the resulting RNA duplex. Double stranded RNA acts as enzyme-dependent antisense through the RNAi/siRNA pathway, involving target mRNA recognition through sense-antisense strand pairing followed by target mRNA degradation by the RNA-induced silencing complex (RISC). Steric blocking antisense (RNase-H-independent antisense) interferes with gene expression or other mRNA-dependent cellular processes by binding to a target sequence of mRNA and getting in the way of other processes. Steric blocking antisense includes 2'-O alkyl (usually in chimeras with RNase-H dependent antisense), peptide nucleic acid (PNA), locked nucleic acid (LNA), and Morpholino antisense.

Antisense mRNA

Antisense RNA is a single-stranded RNA that is complementary to a messenger RNA (mRNA) strand transcribed within a cell. Antisense RNA may be introduced into a cell to inhibit translation of a complementary mRNA by base pairing to it and physically obstructing the translation machinery. This effect is therefore stoichiometric. An example of naturally occurring mRNA antisense mechanism is the hok/sok system of the E. coli R1 plasmid. Antisense RNA has long been thought of as a promising technique for disease therapy; the only such case to have reached the market is the drug fomivirsen. One commentator has characterized antisense RNA as one of "dozens of technologies that are gorgeous in concept, but exasperating in [commercialization]". Generally, antisense RNA still lack effective design, biological activity, and efficient route of administration. Historically, the effects of antisense RNA have often been confused with the effects of RNA interference (RNAi), a related process in which double-stranded RNA fragments called small interfering RNAs trigger catalytically mediated gene silencing, most typically by targeting the RNA-induced silencing complex (RISC) to bind to and degrade the mRNA.

Attempts to genetically engineer transgenic plants to express antisense RNA instead activate the RNAi pathway, although the processes result in differing magnitudes of the same downstream effect, gene silencing. Well-known examples include the Flavr Savr tomato and two cultivars of ringspot-resistant papaya. Transcription of longer cis-antisense transcripts is a common phenomenon in the mammalian transcriptome. Although the function of some cases have been described, such as the Zeb2/Sip1 antisense RNA, no general function has been elucidated. In the case of Zeb2/Sip1, the antisense noncoding RNA is opposite the 5' splice site of an intron in the 5'UTR of the Zeb2 mRNA. Expression of the antisense ncRNA prevents splicing of an intron that contains a ribosome entry site necessary for efficient expression of the Zeb2 protein. Transcription of long antisense ncRNAs is often concordant with the associated protein-coding gene, but more detailed studies have revealed that the relative expression patterns of the mRNA and antisense ncRNA are complex.

Cis-natural Antisense Transcript

Natural antisense transcripts (NATs) are a group of RNAs encoded within a cell that have transcript complementarity to other RNA transcripts. They have been identified in multiple eukaryotes, including humans, mice, yeast and Arabidopsis thaliana. This class of RNAs includes both protein-coding and non-coding RNAs. Current evidence has suggested a variety of regulatory roles for NATs, such as RNA interference (RNAi), alternative splicing, genomic imprinting, and X-chromosome inactivation. NATs are broadly grouped into two categories based on whether they act in cis or in trans. Trans-NATs are transcribed from a different location than their targets and usually have complementarity to multiple transcripts with some mismatches. MicroRNAs (miRNA) are an example of trans-NATs that can target multiple transcripts with a few mismatches. Cis-natural antisense transcripts (cis-NATs) on the other hand are transcribed from the same genomic locus as their target but from the opposite DNA strand and form perfect pairs.

Orientation of cis-NATs

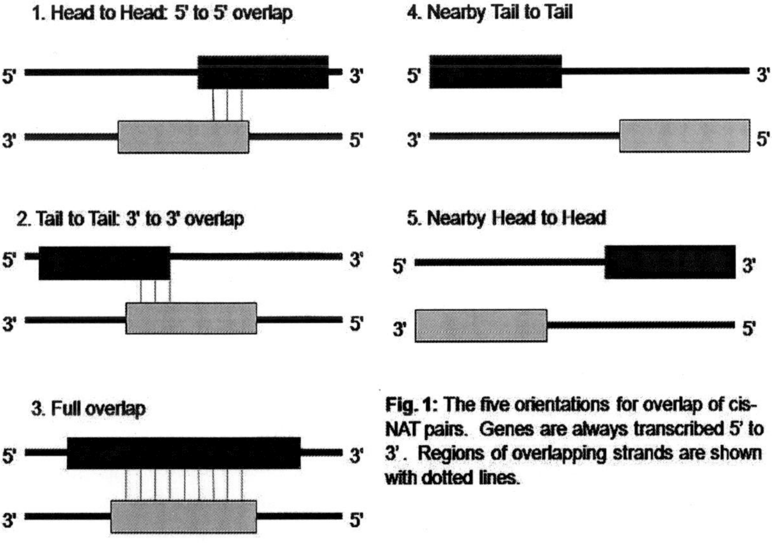

Fig. 1: The five orientations for overlap of cis-NAT pairs. Genes are always transcribed 5' to 3'. Regions of overlapping strands are shown with dotted lines.

Figure: *Orientations of cis-NATs within the genome*

Cis-NATs have a variety of orientations and differing lengths of overlap between pairs. There have been five identified orientations for cis-NATs to date. The most common orientation is head-to-head, where the 5' ends of both transcripts align together. This orientation would result in the greatest knockdown of gene expression if transcriptional collision is the reason for transcript inhibition.

There is however some studies that have suggested that tail-to-tail orientations are the most common NAT pairs. Others such as tail to tail, overlapping, nearby head-to- head, and nearby tail-to-tail are less frequently encountered. Completely overlapping NATs involve the antisense gene being located completely over top of each other . Nearby head-to-head and tail-to-tail orientations are physically discrete from each other but are located very close to each other. Current evidence suggests that there is an overrepresentation of NAT pairs in genes that have catalytic activity. There may be something about these genes in particular that makes them more prone to this type of regulation.

Identification Approach

Identification of NATs in whole genomes is possible due to the large collection of sequence data available from multiple organisms. In silico methods for detecting NATs suffer from several shortcomings depending on the source of sequence information. Studies that use mRNA have sequences whose orientations are known, but the amount of mRNA sequence information available is small.

Predicted gene models using algorithms trained to look for genes gives an increased coverage of the genome at the cost of confidence in the identified gene. Another resource is the extensive expressed sequence tag (EST) libraries but these small sequences must first be assigned an orientation before useful information can be extracted from them.

Some studies have utilized special sequence information in the ESTs such as the poly(A) signal, poly(A) tail, and splicing sites to both filter the ESTs and to give them the correct transcriptional orientation. Combinations of the different sequence sources attempts to maximize coverage as well as maintain integrity in the data.

Pairs of NATs are identified when they form overlapping clusters. There is variability in the cut-off values used in different studies but generally ~20 nucleotides of sequence overlap is considered the minimum for transcripts to be considered and overlapping cluster. Also, transcripts must map to only one other mRNA molecule in order for it to be considered a NAT pair. Currently there are a variety of web and software resources that can be used to look for antisense pairs. The NATsdb or Natural Antisense Transcript database is a rich tool for searching for antisense pairs from multiple organisms.

Mechanisms

Molecular mechanisms behind the regulatory role of cis-NATs are not currently well understood. Three models have been proposed to explain the regulatory effects that cis-NATs have on gene expression. The first model attributes that base pairing between the cis-NAT and its complementary transcript result in a knockdown of mRNA expression.

The assumption of this model is that there will be a precise alignment of at least 6 base pairs between the cis-NAT pair to make double stranded RNA. Epigenetic modifications like DNA methylation and post-translational modification of core histones form the basis of the second model. Although it is not yet clearly understood, it is thought that the reverse transcript guides methylation complexes and/or histone-modifying complexes to the promoter regions of the sense transcript and cause an inhibition of expression from the gene.

Currently it is not known what attributes of cis-NATs are crucial for the epigenetic model of regulation. The final proposed model that has gained favour due to recent experimental evidence is the transcriptional collision model. During the process of transcription of cis-NATs, the transcriptional complexes assemble in the promoter regions of the gene.

RNA polymerases will then begin transcribing the gene at the transcription initiation site laying down nucleotides in a 5' to 3' direction. In the areas of overlap between the cis-NATs the RNA polymerases will collide and stop at the crash site. Transcription is inhibited because RNA polymerases prematurely stop and their incomplete transcripts get degraded.

Importance

Regulation of many biological processes such as development, metabolism and many others requires a careful co-ordination between many different genes; this is usually referred to as a gene regulatory network. A flurry of interest in gene regulatory networks has been sparked by the advent of sequenced genomes of multiple organisms. The next step is to use this information to figure out how genes work together and not just in isolation. During the processes of mammalian development, there is an inactivation of the extra X-chromosome in females. It has been shown that a NAT pair called Xist and Tsix are involved in the hypermethylation of the chromosome. As much as 20-30% of mammalian genes have been shown to be the targets of miRNAs, which highlights the importance of these molecules as regulators across a wide number of genes. Evolutionary reasons for utilizing RNA for regulation of genes may be that it is less costly and faster than synthesizing proteins not needed by the cell. This could have had a selective advantage for early eukaryotes with this type of transcriptional regulation.

Cis-NATs and Disease

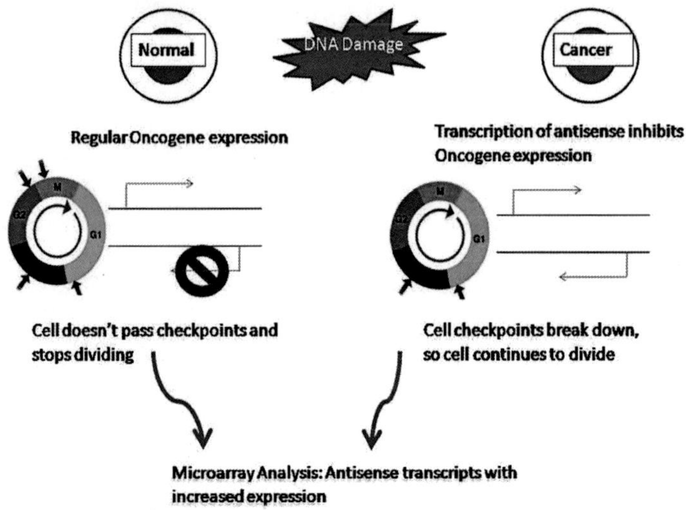

Figure: *Aberrant transcription of antisense transcripts can result in inhibition of oncogenes and allow cell to continue past cell cycle check points. Putative new oncogenes and tumor suppressor genes can be found by looking for upregulated antisense transcripts in cancer cells.*

Antisense transcription might contribute to disease through chromosomal changes that result in the production of aberrant antisense transcripts. A documented case of cis-NATs being involved in human disease comes from an inherited form of á-thalassemia where there is silencing of the hemoglobin á-2 gene through the action of a cis-NAT.

It is thought that in malignant cancer cells with activated transposable elements creates a large amount of transcriptional noise. It is likely that aberrant antisense RNA transcripts resulting from this transcriptional noise may cause stochastic methylation of CpG islands associated with oncogenes and tumor suppressor genes.

This inhibition would further progress the malignancy of the cells since they lose key regulator genes. By looking at upregulated antisense transcripts in tumor cells, researchers and look for more candidate tumor suppressor genes. Also, aberrant cis-NATs have been implicated in neurological diseases such as Parkinson's disease.

Morpholino

In molecular biology, a Morpholino is a molecule in a particular structural family that is used to modify gene expression. Morpholino oligomers (oligos) are an antisense technology used to block access of other molecules to specific sequences within nucleic acid. Morpholinos block small (~25 base) regions of the base-pairing surfaces of ribonucleic acid (RNA).

This article discusses only the Morpholino antisense oligos, which are nucleic acid analogs. The word "morpholino" can occur in other chemical names, referring to chemicals containing a six-membered morpholine ring. To help avoid confusion with other morpholine-containing molecules, when describing oligos "Morpholino" is often capitalized as a trade name, though this usage is not consistent throughout the scientific literature. Morpholinos are sometimes referred to as "PMO" (phosphorodiamidate morpholino oligo). Morpholinos are usually used as a research tool for reverse genetics by knocking down gene function. This is achieved by preventing cells from making a targeted protein or by modifying the splicing of pre-mRNA.

Knocking down gene expression is a powerful method for learning about the function of a particular protein; in similar manner, causing a specific exon to be spliced out of a protein can help to determine the function of the protein moiety encoded by that exon. These molecules have been applied to studies in several model organisms, including mice, zebrafish, frogs, and sea urchins. Morpholinos are also in development as pharmaceutical therapeutics targeted against pathogenic organisms such as bacteria or viruses and for amelioration of genetic diseases. These synthetic oligos were conceived by James E. Summerton (Gene Tools, LLC) and developed in collaboration with Dwight D. Weller (AVI BioPharma Inc.).

Structure

Morpholinos are synthetic molecules that are the product of a redesign of natural nucleic acid structure. Usually 25 bases in length, they bind to complementary sequences of RNA by standard nucleic acid base-pairing. Structurally, the difference between Morpholinos and DNA is that while Morpholinos have standard nucleic acid bases, those bases are bound to morpholine rings instead of deoxyribose rings and linked through phosphorodiamidate groups instead of phosphates. This may be easiest to visualize by referring to the first figure and comparing the structures of the two strands depicted there, one of RNA and the other of a Morpholino. Replacement of anionic phosphates with the uncharged phosphorodiamidate groups eliminates ionization in the usual physiological pH range, so Morpholinos in organisms or cells are uncharged molecules. The entire backbone of a Morpholino is made from these modified subunits. Morpholinos are most commonly used as single-stranded oligos, though heteroduplexes of a Morpholino strand and a complementary DNA strand may be used in combination with cationic cytosolic delivery reagents.

Function

Morpholinos do not degrade their target RNA molecules, unlike many antisense structural types (e.g., phosphorothioates, siRNA). Instead, Morpholinos act by "steric blocking", binding to a target sequence within an RNA and simply getting in the way of molecules that might otherwise interact with the RNA. Morpholino oligos are often used to investigate the role of a specific mRNA transcript in an embryo. Developmental biologists inject Morpholino oligos into eggs or embryos of zebrafish, African clawed frog (Xenopus), sea urchin, and killifish (F. heteroclitus) producing morphant embryos, or electroporate Morpholinos into chick embryos at later stages of development. With appropriate cytosolic delivery systems, Morpholinos are effective in cell culture. Morpholinos are being developed as pharmaceuticals under the name "NeuGenes" by AVI BioPharma Inc. They have been used in mammals such as mice and some are currently being tested in clinical trials for treatment of Duchenne muscular dystrophy in humans.

Normal Gene Expression in Eukaryotes

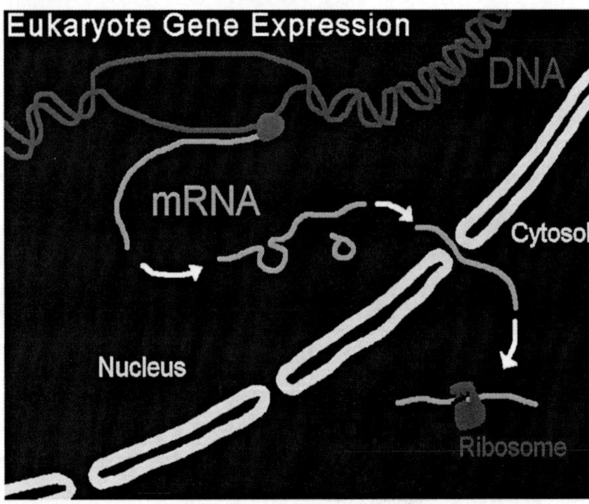

***Figure:** Eukaryotic gene expression without intervention by a Morpholino*

In eukaryotic organisms, pre-mRNA is transcribed in the nucleus, introns are spliced out, then the mature mRNA is exported from the nucleus to the cytoplasm. The small subunit of the ribosome usually starts by binding to one end of the mRNA and is joined there by various other eukaryotic initiation factors, forming the initiation complex. The initiation complex scans along the mRNA strand until it reaches a start codon, and then the large subunit of the ribosome attaches to the small subunit and translation of a protein begins. This entire process is referred to as gene expression; it is the process by which the information in a gene, encoded as a sequence of bases in DNA, is converted into the structure of a protein. A Morpholino can modify splicing or block translation, depending on the Morpholino's base sequence.

Blocking Translation

Bound to the 5'-untranslated region of messenger RNA (mRNA), Morpholinos can interfere with progression of the ribosomal initiation complex from the 5' cap to the start codon. This prevents translation of the coding region of the targeted transcript (called "knocking down" gene expression).

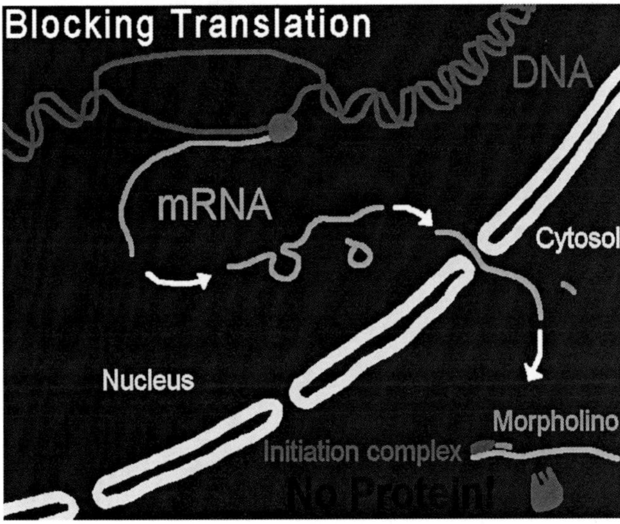

Figure: *Translation blocked by a Morpholino oligo.*

This is useful experimentally when an investigator wishes to know the function of a particular protein; Morpholinos provide a convenient means of knocking down expression of the protein and learning how that knockdown changes the cells or organism. Some Morpholinos knock down expression so effectively that, after degradation of preexisting proteins, the targeted proteins become undetectable by Western blot.

Modifying pre-mRNA Splicing

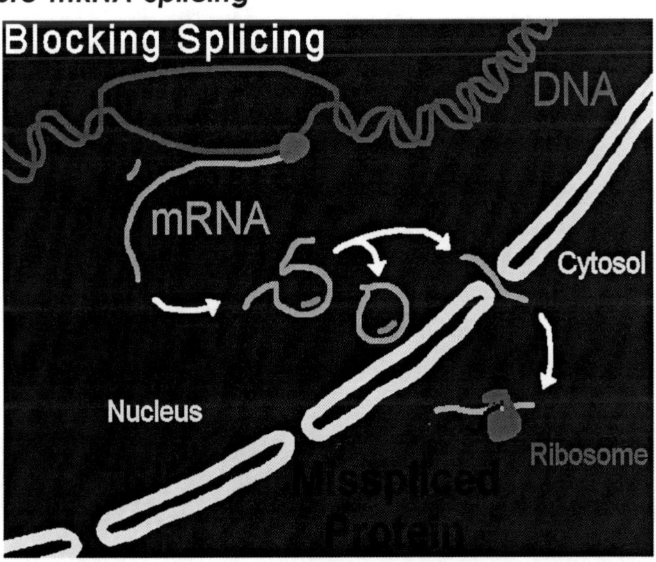

Figure: *Splicing blocked by a Morpholino oligo*

Morpholinos can interfere with pre-mRNA processing steps either by preventing splice-directing small nuclear ribonucleoproteins (snRNP) complexes from binding to their targets at the borders of introns on a strand of pre-mRNA, or by blocking the nucleophilic adenine base and preventing it from forming the splice lariat structure, or by interfering with the binding of splice regulatory proteins such as splice silencers and splice enhancers.

Preventing the binding of snRNP U1 (at the donor site) or U2/U5 (at the polypyrimidine moiety and acceptor site) can cause modified splicing, commonly excluding exons from the mature mRNA. Targeting some splice targets results in intron inclusions, while activation of cryptic splice sites can lead to partial inclusions or exclusions. Targets of U11/U12 snRNPs can also be blocked. Splice modification can be conveniently assayed by reverse-transcriptase polymerase chain reaction (RT-PCR) and is seen as a band shift after gel electrophoresis of RT-PCR products.

Blocking Other mRNA Sites

Morpholinos have been used to block miRNA activity and maturation. They can also block ribozyme activity. U2 and U12 snRNP functions have been inhibited by Morpholinos. Morpholinos targeted to "slippery" mRNA sequences within protein coding regions can induce translational frameshifts. Activities of Morpholinos against this variety of targets suggest that Morpholinos can be used as a general-purpose tool for blocking interactions of proteins or nucleic acids with mRNA.

Specificity, Stability and Non-antisense Effects

Morpholinos have become a standard knockdown tool in animal embryonic systems, which have a broader range of gene expression than adult cells and can be strongly affected by an off-target interaction. Following initial injections into frog or fish embryos at the single-cell or few-cell stages, Morpholino effects can be measured up to five days later, after most of the processes of organogenesis and differentiation are past, with observed phenotypes consistent with target-gene knockdown. Control oligos with irrelevant sequences usually produce no change in embryonic phenotype, evidence of the Morpholino oligo's sequence-specificity and lack of non-antisense effects. The dose required for a knockdown can be reduced by coinjection of several Morpholino oligos targeting the same mRNA, which is an effective strategy for reducing or eliminating dose-dependent off-target RNA interactions.

mRNA rescue experiments can often restore the wild-type phenotype to the embryos and provide evidence for the specificity of a Morpholino. In an mRNA rescue, a Morpholino is co-injected with an mRNA that codes for the same protein that the Morphlino is targeted to knock down. However, the rescue mRNA has a modified 5'-UTR (untranslated region) so that the rescue mRNA contains no target for the Morpholino but the rescue mRNA's coding region encodes the protein of interest. Translation of the rescue mRNA replaces production of the protein that was knocked down by the Morpholino. Since the rescue mRNA would not affect phenotypic changes due to modulation of off-target gene expression by the Morpholino, this return to wild-type phenotype is further evidence of Morpholino specificity.

Because of their completely unnatural backbones, Morpholinos are not recognized by cellular proteins. Nucleases do not degrade Morpholinos, nor are they degraded in serum or in cells. Morpholinos do not activate toll-like receptors and so they do not activate innate immune responses such as interferon induction or the NF-êB mediated inflammation response. Morpholinos are not known to modify methylation of DNA.

A cause for concern in the use of Morpholinos is the potential for "off-target" effects. Up to 18% of Morpholinos appear to have nontarget-related phenotypes including cell death in the central nervous system and somite tissues of zebrafish embryos. Most of these effects have been shown to be due to activation of p53-mediated apoptosis, and can be suppressed by co-injection of an anti-p53 Morpholino along with the experimental Morpholino; moreover, the p53-mediated apoptotic effect of a Morpholino knockdown has been phenocopied using another antisense structural type, showing the p53-mediated apoptosis to be a consequence of the loss of the targeted protein and not a consequence of the oligo type used to knock down the protein. It appears that these effects are sequence-specific; as in most cases, if a Morpholino is associated with non-target effects, the 4-base mismatch Morpholino will not trigger these effects.

The question of whether an observed morphant phenotype is due to the intended knockdown or an off-target interaction can often be addressed by running another experiment to confirm that the observed morphant phenotype results from the knockdown of the expected target. This can be done by recapitulating the morphant phenotype with a second, non-overlapping Morpholino targeting the same mRNA or by confirmation of the observed phenotypes by use of a mutant strain or dominant-negative methods. As mentioned above, rescue of observed phenotypes by coinjecting a rescue mRNA is, when feasible, a reliable test of specificity of a Morpholino.

Delivery

For a Morpholino to be effective, it must be delivered past the cell membrane into the cytosol of a cell. Once in the cytosol, Morpholinos freely diffuse between the cytosol and nucleus, as demonstrated by the nuclear splice-modifying activity of Morpholinos observed after microinjection into the cytosol of cells. Different methods are used for delivery into embryos, into cultured cells or into adult animals. A microinjection apparatus is usually used for delivery into an embryo, with injections most commonly performed at the single-cell or few-cell stage; an alternative method for embryonic delivery is electroporation, which can deliver oligos into tissues of later embryonic stages. Common techniques for delivery into cultured cells include the Endo-Porter peptide (which causes the Morpholino to be released from endosomes), the Special Delivery system (using a Morpholino-DNA heteroduplex and an ethoxylated polyethylenimine delivery reagent), electroporation, or scrape loading. Delivery into adult tissues is usually difficult, though there are a few systems allowing useful uptake of unmodified Morpholino oligos (including the inherently leaky muscle cells caused by Duchenne muscular dystrophy or the vascular endothelial cells stressed during balloon angioplasty).

Though they permeate through intercellular spaces in tissues effectively, unconjugated PMOs have limited distribution into the cytosol and nuclear spaces within healthy tissues following IV administration. Systemic delivery into many cells in adult organisms can be accomplished by using covalent conjugates of Morpholino oligos with cell-penetrating peptides, and, while toxicity has been associated with moderate doses of the peptide conjugates, they have been used in vivo for effective oligo delivery at doses below those causing observed toxicity. An octa-guanidinium dendrimer attached to the end of a Morpholino can deliver the modified oligo (called a Vivo-Morpholino) from the blood to the cytosol. Delivery-enabled Morpholinos, such as peptide conjugates and Vivo-Morpholinos, show promise as therapeutics for viral and genetic diseases.

Peptide Nucleic Acid

Peptide nucleic acid (PNA) is an artificially synthesized polymer similar to DNA or RNA invented by Peter E. Nielsen (Univ. Copenhagen), Michael Egholm (Univ. Copenhagen), Rolf H. Berg (Risø National Lab), and Ole Buchardt (Univ. Copenhagen) in 1991.

DNA and RNA have a deoxyribose and ribose sugar backbone, respectively, whereas PNA's backbone is composed of repeating N-(2-aminoethyl)-glycine units linked by peptide bonds. PNA is not known to occur naturally. The various purine and pyrimidine bases are linked to the backbone by methylene carbonyl bonds. PNAs are depicted like peptides, with the N-terminus at the first (left) position and the C-terminus at the right.

Since the backbone of PNA contains no charged phosphate groups, the binding between PNA/DNA strands is stronger than between DNA/DNA strands due to the lack of electrostatic repulsion. Unfortunately, this also causes it to be rather hydrophobic, which makes it difficult to deliver to body cells in solution without being flushed out of the body first. Early experiments with homopyrimidine strands (strands consisting of only one repeated pyrimidine base) have shown that the T_m ("melting" temperature) of a 6-base thymine PNA/adenine DNA double helix was 31 °C in comparison to an equivalent 6-base DNA/DNA duplex that denatures at a temperature less than 10 °C. Mixed base PNA molecules are true mimics of DNA molecules in terms of base-pair recognition. PNA/PNA binding is stronger than PNA/DNA binding.

Synthetic peptide nucleic acid oligomers have been used in recent years in molecular biology procedures, diagnostic assays and antisense therapies. Due to their higher binding strength it is not necessary to design long PNA oligomers for use in these roles, which usually require oligonucleotide probes of 20–25 bases. The main concern of the length of the PNA-oligomers is to guarantee the specificity. PNA oligomers also show greater specificity in binding to complementary DNAs, with a PNA/DNA base mismatch being more destabilizing than a similar mismatch in a DNA/DNA duplex. This binding strength and specificity also applies to PNA/RNA duplexes. PNAs are not easily recognized by either nucleases or proteases, making them resistant to enzyme degradation. PNAs are also stable over a wide pH range. Though an unmodified PNA cannot readily cross cell membranes to enter the cytosol, covalently coupling a cell penetrating peptide to a PNA can improve cytosolic delivery.

It has been hypothesized that the earliest life on Earth may have used PNA as a genetic material due to its extreme robustness, simpler formation and possible spontaneous polymerization at 100°C (while water at standard pressure boils at this temperature, water at high pressure—as in deep ocean—boils at higher temperatures). If this is so, life evolved to a DNA/RNA-based system only at a later stage. Evidence for this PNA world hypothesis is however far from conclusive. The name is somewhat of a misnomer as PNA is not an acid.

Applications

- Alteration of gene expression - both as inhibitor and promoter in different cases
- Antigene and antisense therapeutic agent
- Anticancer agent
- Antiviral, Antibacterial and Antiparasitic agent
- Molecular tools and probes of biosensor
- Detection of DNA sequences
- Locked Nucleic Acid

A locked nucleic acid (LNA), often referred to as inaccessible RNA, is a modified RNA nucleotide. The ribose moiety of an LNA nucleotide is modified with an extra bridge connecting the 2' oxygen and 4' carbon. The bridge "locks" the ribose in the 3'-endo (North) conformation, which is often found in the A-form duplexes. LNA nucleotides can be mixed with DNA or RNA residues in the oligonucleotide whenever desired. Such oligomers are synthesized chemically and are commercially available. The locked ribose conformation enhances base stacking and backbone pre-organization. This significantly increases the hybridization properties (melting temperature) of oligonucleotides. LNA was independently synthesized by the group of Jesper Wengel in 1998, soon after the first synthesis by the group of Takeshi Imanishi in 1997. The exclusive rights to the LNA technology were secured in 1997 by Exiqon A/S, a Danish biotech company.

LNA nucleotides are used to increase the sensitivity and specificity of expression in DNA microarrays, FISH probes, real-time PCR probes and other molecular biology techniques based on oligonucleotides. For the in situ detection of miRNA the use of LNA is currently (2005) the only efficient method. A triplet of LNA nucleotides surrounding a single-base mismatch site maximizes LNA probe specificity unless the probe contains the guanine base of G-T mismatch. Using LNA based oligonucleotides therapeutically is an emerging field in biotechnology.

The Danish pharmaceutical company Santaris Pharma a/s owns the sole rights to therapeutic uses of LNA technology, and is now developing a new, LNA based, hepatitis C drug called miravirsen, targeting miR-122, which is in Phase II clinical testing as of late 2010.

Benefits of the LNA technology

Some of the benefits of using LNA include:

- Ideal for the detection of short RNA and DNA targets
- Increases the thermal stability of duplexes
- Capable of single nucleotide discrimination
- Resistant to exo- and endonucleases resulting in high stability in vivo and in vitro applications
- Increased target specificity
- Facilitates Tm normalization
- Strand invasion properties enables detection of "hard to access" samples
- Compatible with standard enzymatic processes

Applications of the LNA technology

Some proven applications of LNA include:

- Allele-specific PCR: allows for the design of shorter primers, without compromising binding specificity
- Microarray gene expression profiling: provides increased sensitivity and selectivity with smaller amounts of substrates

- Small RNA research
- SNP genotyping
- mRNA antisense oligonucleotides
- RNAi
- DNAzymes
- Fluorescence Polarization probes
- Molecular Beacons
- Gene repair/exon skipping
- Splice variant detection
- Comparative genome hybridization (GCH)

Other therapeutic and diagnostic applications of LNA technology are in development.

RNA interference (which uses double-strand RNA)

RNA interference (RNAi) is a system within living cells that takes part in controlling which genes are active and how active they are. Two types of small RNA molecules – microRNA (miRNA) and small interfering RNA (siRNA) – are central to RNA interference. RNAs are the direct products of genes, and these small RNAs can bind to other specific RNAs (mRNA) and either increase or decrease their activity, for example by preventing a messenger RNA from producing a protein. RNA interference has an important role in defending cells against parasitic genes – viruses and transposons – but also in directing development as well as gene expression in general.

The RNAi pathway is found in many eukaryotes including animals and is initiated by the enzyme Dicer, which cleaves long double-stranded RNA (dsRNA) molecules into short fragments of ~20 nucleotides that are called siRNAs. Each siRNA is unwound into two single-stranded (ss) ssRNAs, namely the passenger strand and the guide strand. The passenger strand will be degraded, and the guide strand is incorporated into the RNA-induced silencing complex (RISC). The most well-studied outcome is post-transcriptional gene silencing, which occurs when the guide strand base pairs with a complementary sequence of a messenger RNA molecule and induces cleavage by Argonaute, the catalytic component of the RISC complex. This process is known to spread systemically throughout the organism despite initially limited molar concentrations of siRNA.

The selective and robust effect of RNAi on gene expression makes it a valuable research tool, both in cell culture and in living organisms because synthetic dsRNA introduced into cells can induce suppression of specific genes of interest. RNAi may also be used for large-scale screens that systematically shut down each gene in the cell, which can help identify the components necessary for a particular cellular process or an event such as cell division. Exploitation of the pathway is also a promising tool in biotechnology and medicine. Historically, RNA interference was known by other names, including cosuppression, post transcriptional gene silencing, and quelling. Only after these apparently unrelated processes were fully understood did it become clear that they all described the RNAi phenomenon. In 2006, Andrew Fire and Craig C. Mello shared the Nobel Prize in Physiology or Medicine for their work on RNA interference in the nematode worm C. elegans, which they published in 1998.

Cellular Mechanism

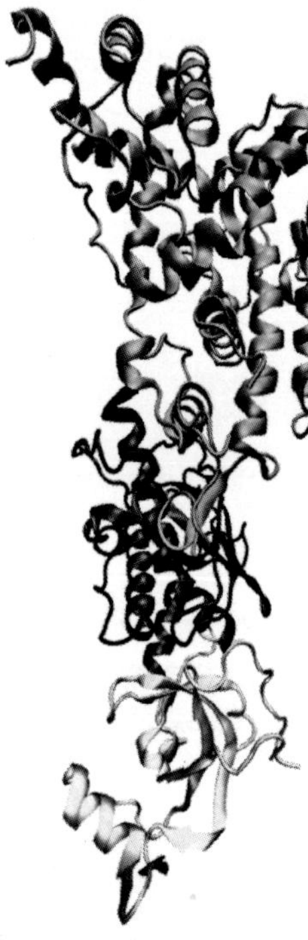

The dicer protein from Giardia intestinalis, which catalyzes the cleavage of dsRNA to siRNAs. The RNase domains are colored green, the PAZ domain yellow, the platform domain red, and the connector helix blue. RNAi is an RNA-dependent gene silencing process that is controlled by the RNA-induced silencing complex (RISC) and is initiated by short double-stranded RNA molecules in a cell's cytoplasm, where they interact with the catalytic RISC component argonaute. When the dsRNA is exogenous (coming from infection by a virus with an RNA genome or laboratory manipulations), the RNA is imported directly into the cytoplasm and cleaved to short fragments by the enzyme. The initiating dsRNA can also be endogenous (originating in the cell), as in pre-microRNAs expressed from RNA-coding genes in the genome. The primary transcripts from such genes are first processed to form the characteristic stem-loop structure of pre-miRNA in the nucleus, then exported to the cytoplasm to be cleaved by Dicer. Thus, the two dsRNA pathways, exogenous and endogenous, converge at the RISC complex.

dsRNA Cleavage

Endogenous dsRNA initiates RNAi by activating the ribonuclease protein Dicer, which binds and cleaves double-stranded RNAs (dsRNAs) to produce double-stranded fragments of 20–25 base pairs with a 2-nucleotide overhang at the 3' end. Bioinformatics studies on the genomes of multiple organisms suggest this length maximizes target-gene specificity and minimizes non-specific effects.

These short double-stranded fragments are called small interfering RNAs (siRNAs). These siRNAs are then separated into single strands and integrated into an active RISC complex. After integration into the RISC, siRNAs base-pair to their target mRNA and induce cleavage of the mRNA, thereby preventing it from being used as a translation template. Exogenous dsRNA is detected and bound by an effector protein, known as RDE-4 in C. elegans and R2D2 in Drosophila, that stimulates dicer activity. This protein only binds long dsRNAs, but the mechanism producing this length specificity is unknown. This RNA-binding protein then facilitates the transfer of cleaved siRNAs to the RISC complex. In C. elegans, this initiation response is amplified through the synthesis of a population of 'secondary' siRNAs during which the dicer-produced initiating or 'primary' siRNAs are used as templates. These 'secondary' siRNAs are structurally distinct from dicer-produced siRNAs and appear to be produced by an RNA-dependent RNA polymerase (RdRP).

MicroRNA

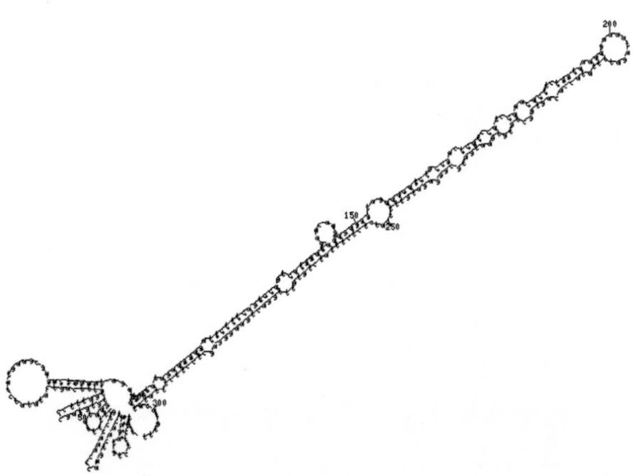

Figure: *The stem-loop secondary structure of a pre-microRNA from Brassica oleracea.*

MicroRNAs (miRNAs) are genomically encoded non-coding RNAs that help regulate gene expression, particularly during development. The phenomenon of RNA interference, broadly defined, includes the endogenously induced gene silencing effects of miRNAs as well as silencing triggered by foreign dsRNA. Mature miRNAs are structurally similar to siRNAs produced from exogenous dsRNA, but before reaching maturity, miRNAs must first undergo extensive post-transcriptional modification. An miRNA is expressed from a much longer RNA-coding gene as a primary transcript known as a pri-miRNA which is processed, in the cell nucleus, to a 70-nucleotide stem-loop structure called a pre-miRNA by the microprocessor complex.

This complex consists of an RNase III enzyme called Drosha and a dsRNA-binding protein Pasha. The dsRNA portion of this pre-miRNA is bound and cleaved by Dicer to produce the mature miRNA molecule that can be integrated into the RISC complex; thus, miRNA and siRNA share the same cellular machinery downstream of their initial processing. The siRNAs derived from long dsRNA precursors differ from miRNAs in that miRNAs, especially those in animals, typically have incomplete base pairing to a target and inhibit the translation of many different mRNAs with similar sequences. In contrast, siRNAs typically base-pair perfectly and induce mRNA cleavage only in a single, specific target. In Drosophila and C. elegans, miRNA and siRNA are processed by distinct argonaute proteins and dicer enzymes.

***Figure:** Left: A full-length argonaute protein from the archaea species Pyrococcus furiosus. Right: The PIWI domain of an argonaute protein in complex with double-stranded RNA.*

RISC Activation and Catalysis

The active components of an RNA-induced silencing complex (RISC) are endonucleases called argonaute proteins, which cleave the target mRNA strand complementary to their bound siRNA. As the fragments produced by dicer are double-stranded, they could each in theory produce a functional siRNA. However, only one of the two strands, which is known as the guide strand, binds the argonaute protein and directs gene silencing.

The other anti-guide strand or passenger strand is degraded during RISC activation. Although it was first believed that an ATP-dependent helicase separated these two strands, the process is actually ATP-independent and performed directly by the protein components of RISC. The strand selected as the guide tends to be the one whose 5' end is less stably paired to its complement, but strand selection is unaffected by the direction in which dicer cleaves the dsRNA before RISC incorporation. Instead, the R2D2 protein may serve as the differentiating factor by binding the more-stable 5' end of the passenger strand.

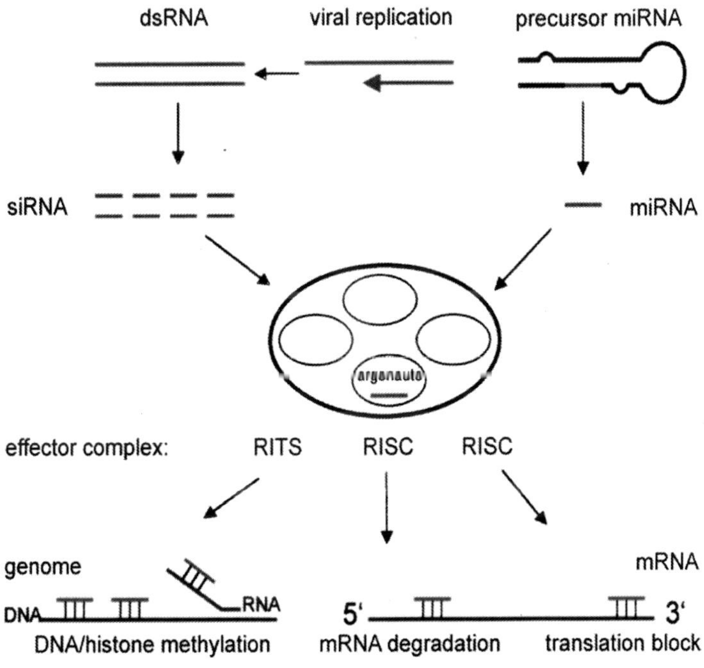

The structural basis for binding of RNA to the argonaute protein was examined by X-ray crystallography of the binding domain of an RNA-bound argonaute protein. Here, the phosphorylated 5' end of the RNA strand enters a conserved basic surface pocket and makes contacts through a divalent cation (an atom with two positive charges) such as magnesium and by aromatic stacking (a process that allows more than one atom to share an electron by passing it back and forth) between the 5' nucleotide in the siRNA and a conserved tyrosine residue. This site is thought to form a nucleation site for the binding of the siRNA to its mRNA target.

It is not understood how the activated RISC complex locates complementary mRNAs within the cell. Although the cleavage process has been proposed to be linked to translation, translation of the mRNA target is not essential for RNAi-mediated degradation. Indeed, RNAi may be more effective against mRNA targets that are not translated. Argonaute proteins, the catalytic components of RISC, are localized to specific regions in the cytoplasm called P-bodies (also cytoplasmic bodies or GW bodies), which are regions with high rates of mRNA decay; miRNA activity is also clustered in P-bodies. Disruption of P-bodies decreases the efficiency of RNA interference, suggesting that they are the site of a critical step in the RNAi process. The enzyme dicer trims double stranded RNA, to form small interfering RNA or microRNA. These processed RNAs are incorporated into the RNA-induced silencing complex (RISC), which targets messenger RNA to prevent translation.

Transcriptional Silencing

Components of the RNA interference pathway are also used in many eukaryotes in the maintenance of the organization and structure of their genomes. Modification of histones and associated induction of heterochromatin formation serves to downregulate genes pre-transcriptionally; this process is referred to as RNA-induced transcriptional silencing (RITS), and is carried out by a complex of proteins called the RITS complex. In fission yeast this complex contains argonaute, a chromodomain protein Chp1, and a protein called Tas3 of unknown function. As a consequence, the induction and spread of heterochromatic regions requires the argonaute and RdRP proteins. Indeed, deletion of these genes in the fission yeast S. pombe disrupts histone methylation and centromere formation, causing slow or stalled anaphase during cell division. In some cases, similar processes associated with histone modification have been observed to transcriptionally upregulate genes.

The mechanism by which the RITS complex induces heterochromatin formation and organization is not well understood, and most studies have focused on the mating-type region in fission yeast, which may not be representative of activities in other genomic regions or organisms. In maintenance of existing heterochromatin regions, RITS forms a complex with siRNAs complementary to the local genes and stably binds local methylated histones, acting co-transcriptionally to degrade any nascent pre-mRNA transcripts that are initiated by RNA polymerase. The formation of such a heterochromatin region, though not its maintenance, is dicer-dependent, presumably because dicer is required to generate the initial complement of siRNAs that target subsequent transcripts. Heterochromatin maintenance has been suggested to function as a self-reinforcing feedback loop, as new siRNAs are formed from the occasional nascent transcripts by RdRP for incorporation into local RITS complexes. The relevance of observations from fission yeast mating-type regions and centromeres to mammals is not clear, as heterochromatin maintenance in mammalian cells may be independent of the components of the RNAi pathway.

Crosstalk with RNA Editing

The type of RNA editing that is most prevalent in higher eukaryotes converts adenosine nucleotides into inosine in dsRNAs via the enzyme adenosine deaminase (ADAR). It was originally proposed in 2000 that the RNAi and A'!I RNA editing pathways might compete for a common dsRNA substrate. Indeed, some pre-miRNAs do undergo A'!I RNA editing, and this mechanism may regulate the processing and expression of mature miRNAs. Furthermore, at least one mammalian ADAR can sequester siRNAs from RNAi pathway components. Further support for this model comes from studies on ADAR-null C. elegans strains indicating that A'!I RNA editing may counteract RNAi silencing of endogenous genes and transgenes.

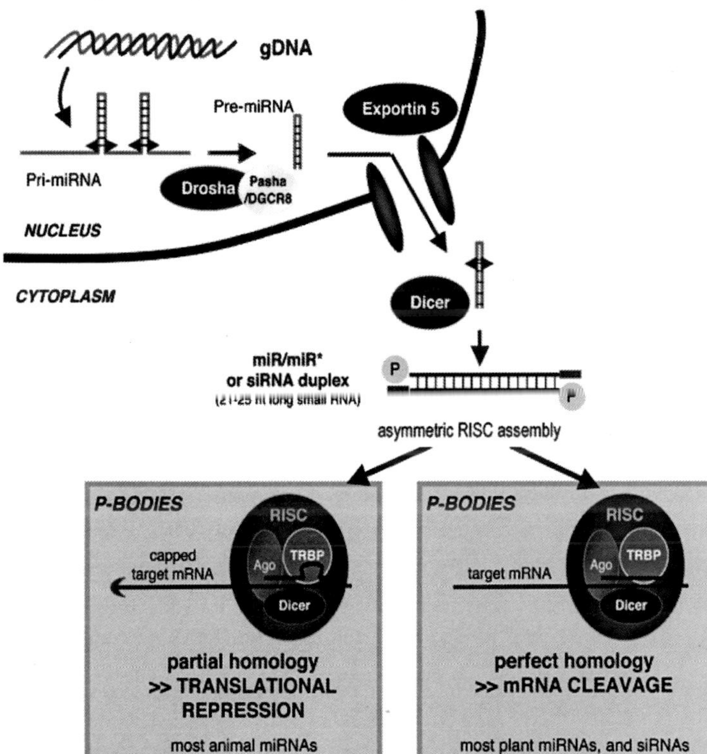

Figure: *Illustration of the major differences between plant and animal gene silencing. Natively expressed microRNA or exogenous small interfering RNA is processed by dicer and integrated into the RISC complex, which mediates gene silencing.*

Variation Among Organisms

Organisms vary in their ability to take up foreign dsRNA and use it in the RNAi pathway. The effects of RNA interference can be both systemic and heritable in plants and C. elegans, although not in Drosophila or mammals.

In plants, RNAi is thought to propagate by the transfer of siRNAs between cells through plasmodesmata (channels in the cell walls that enable communication and transport). The heritability comes from methylation of promoters targeted by RNAi; the new methylation pattern is copied in each new generation of the cell. A broad general distinction between plants and animals lies in the targeting of endogenously produced miRNAs; in plants, miRNAs are usually perfectly or nearly perfectly complementary to their target genes and induce direct mRNA cleavage by RISC, while animals' miRNAs tend to be more divergent in sequence and induce translational repression. This translational effect may be produced by inhibiting the interactions of translation initiation factors with the messenger RNA's polyadenine tail.

Some eukaryotic protozoa such as Leishmania major and Trypanosoma cruzi lack the RNAi pathway entirely. Most or all of the components are also missing in some fungi, most notably the model organism Saccharomyces cerevisiae. A recent study however reveals the presence of RNAi in other budding yeast species such as Saccharomyces castellii and Candida albicans, further demonstrating that inducing two RNAi-related proteins from S. castellii facilitates RNAi in S. cerevisiae. That certain ascomycetes and basidiomycetes are missing RNA interference pathways indicates that proteins required for RNA silencing have been lost independently from many fungal lineages, possibly due to the evolution of a novel pathway with similar function, or to the lack of selective advantage in certain niches.

Related Prokaryotic Systems

Gene expression in prokaryotes is influenced by an RNA-based system similar in some respects to RNAi. Here, RNA-encoding genes control mRNA abundance or translation by producing a complementary RNA that binds to an mRNA by base pairing. However these regulatory RNAs are not generally considered to be analogous to miRNAs because the dicer enzyme is not involved. It has been suggested that CRISPR interference systems in prokaryotes are analogous to eukaryotic RNA interference systems, although none of the protein components are orthologous.

Biological Functions

Immunity

RNA interference is a vital part of the immune response to viruses and other foreign genetic material, especially in plants where it may also prevent the self-propagation of transposons. Plants such as Arabidopsis thaliana express multiple dicer homologs that are specialized to react differently when the plant is exposed to different types of viruses. Even before the RNAi pathway was fully understood, it was known that induced gene silencing in plants could spread throughout the plant in a systemic effect, and could be transferred from stock to scion plants via grafting. This phenomenon has since been recognized as a feature of the plant adaptive immune system, and allows the entire plant to respond to a virus after an initial localized encounter. In response, many plant viruses have evolved elaborate mechanisms that suppress the RNAi response in plant cells. These include viral proteins that bind short double-stranded RNA fragments with single-stranded overhang ends, such as those produced by the action of dicer. Some plant genomes also express endogenous siRNAs in response to infection by specific types of bacteria. These effects may be part of a generalized response to pathogens that downregulates any metabolic processes in the host that aid the infection process. Although animals generally express fewer variants of the dicer enzyme than plants, RNAi in some animals has also been shown to produce an antiviral response.

In both juvenile and adult Drosophila, RNA interference is important in antiviral innate immunity and is active against pathogens such as Drosophila X virus. A similar role in immunity may operate in C. elegans, as argonaute proteins are upregulated in response to viruses and worms that overexpress components of the RNAi pathway are resistant to viral infection.

Downregulation of Genes

Endogenously expressed miRNAs, including both intronic and intergenic miRNAs, are most important in translational repression and in the regulation of development, especially on the timing of morphogenesis and the maintenance of undifferentiated or incompletely differentiated cell types such as stem cells. The role of endogenously expressed miRNA in downregulating gene expression was first described in C. elegans in 1993. In plants this function was discovered when the "JAW microRNA" of Arabidopsis was shown to be involved in the regulation of several genes that control plant shape. In plants, the majority of genes regulated by miRNAs are transcription factors; thus miRNA activity is particularly wide-ranging and regulates entire gene networks during development by modulating the expression of key regulatory genes, including transcription factors as well as F-box proteins. In many organisms, including humans, miRNAs have also been linked to the formation of tumors and dysregulation of the cell cycle. Here, miRNAs can function as both oncogenes and tumor suppressors.

Upregulation of Genes

RNA sequences (siRNA and miRNA) that are complementary to parts of a promoter can increase gene transcription, a phenomenon dubbed RNA activation. Part of the mechanism for how these RNA upregulate genes is known: dicer and argonaute are involved, possibly via histone demethylation. miRNAs have also been proposed to upregulate their target genes upon cell cycle arrest, although the underlying mechanisms have not been elucidated.

Evolution

Based on parsimony-based phylogenetic analysis, the most recent common ancestor of all eukaryotes most likely already possessed an early RNA interference pathway; the absence of the pathway in certain eukaryotes is thought to be a derived characteristic. This ancestral RNAi system probably contained at least one dicer-like protein, one argonaute, one PIWI protein, and an RNA-dependent RNA polymerase that may have also played other cellular roles. A large-scale comparative genomics study likewise indicates that the eukaryotic crown group already possessed these components, which may then have had closer functional associations with generalized RNA degradation systems such as the exosome. This study also suggests that the RNA-binding argonaute protein family, which is shared among eukaryotes, most archaea, and at least some bacteria (such as Aquifex aeolicus), is homologous to and originally evolved from components of the translation initiation system.

The ancestral function of the RNAi system is generally agreed to have been immune defense against exogenous genetic elements such as transposons and viral genomes. Related functions such as histone modification may have already been present in the ancestor of modern eukaryotes, although other functions such as regulation of development by miRNA are thought to have evolved later.

RNA interference genes, as components of the antiviral innate immune system in many eukaryotes, are involved in an evolutionary arms race with viral genes. Some viruses have evolved mechanisms for suppressing the RNAi response in their host cells, an effect that has been noted particularly for plant viruses. Studies of evolutionary rates in Drosophila have shown that genes in the RNAi pathway are subject to strong directional selection and are among the fastest-evolving genes in the Drosophila genome.

5

Functional Genomics

The effects of six glucocorticoids (dexamethasone, hydrocortisone, 6-methylprednisolone, prednisolone,prednisone, and triamcinolone) on inducible gene expression, based on the chimaeric transcriptional activator GVG and carried by the binary expression vector pINDEX3-m-gfp5-ER, were evaluated in transgenic Virginia pine cell cultures. The concentration that activated GVG transcription factor activity, the level of inducible m-gfp5-ER expression, and the kinetics of inducible m-gfp5-ER expression were determined for each glucocorticoid. Transgenic cells produced green fluorescence upon blue light excitation after treatment with prednisolone, prednisone, 6-methylprednisolone, dexamethasone, triamcinolone, and hydrocortisone.

Green fluorescence was observed at 6–12 h after treatment of all six glucocorticoids at concentrations of 1, 3, 5, and 10 mg 121. Differential expression of gfp was confirmed by northern blot analysis and by quantitative fluorescence analyses of confocal images taken by a LSM 510 Laser Scanning Microscope. Fresh and dry weight increases of transgenic cell cultures were not affected by all six glucocorticoids at concentrations of 0.1, 0.5, 1, 3, and 5 mg 121. It is shown that triamcinolone had the most potent effect on the GVG system.

Different glucocorticoids can therefore be used to regulate the GVG transcriptional activator and to induce gene expression in transgenic plant cells, and this property could be useful in establishing an optimum system of transgene regulation. Conditional gene expression systems, based on the interaction of a chemical inducer with a specifically designed transcription factor, which results in the transactivation of a synthetic promoter upstream of the target gene, have proved to be powerful tools for basic research in molecular biology and for biotechnological applications with transgenic animals and plants (Aoyama, 1999; Reynolds, 1999; Saez et al., 1997; Wirtz and Clayton, 1995; Zuo and Chua, 2000).

Inducible gene expression systems, where the expression of target genes are tightly controlled by chemical inducers, are extremely useful for: (i) analysing biochemical and developmental processes (Gatz, 1997; Ward et al., 1993); (ii) determining temporal and spatial expression patterns of specific genes; and (iii) studying transgenes where their constitutive expression may have deleterious effects on transformed organisms (Aoyama and Chua, 1997; Ouwerkerk et al., 2001). Inducible gene expression systems may also be used to analyse gene function, sitespecific DNA excision, activation tagging, conditional genetic complementation, and the restoration of male fertility (Padidam, 2003; Reynolds, 1999).

Several chemical-inducible systems based on activation and inactivation of the target gene have been developed (Gatz, 1997; Padidam, 2003; Reynolds, 1999; Shockett et al., 1995; Zuo and Chua, 2000). Artificial promoters that combine tight transcriptional control and induction by low concentrations of chemical inducers have been developed from bacterial repressor or animal steroid-binding elements (Gatz, 1997; Love et al., 2002; Zuo and Chua, 2000). These promoters include tetracycline-inducible Triple-Op (Gatz and Quail, 1988), tetracycline-repressible Top10 (Hinrichs et al., 1994; Love et al., 2002; Weinmann et al., 1994), glucocorticoid-induced GAL4-UAS (Aoyama and Chua, 1997), ethanol-inducible alcA (Caddick et al., 1998), and ecdysone-inducible GRH (Martinez et al., 1999). Chemicals that are used to regulate transgene expression include the antibiotic tetracycline, the steroids dexamethasone and estradiol, herbicide safeners, copper, ethanol, the inducer of pathogen-related proteins, benzothiadiazol, and the insecticide methoxyfenozide (Padidam, 2003; Reynolds, 1999). In the case of GVG-regulated gene expression (Aoyama and Chua 1997; Ouwerkerk et al., 2001), the chimaeric transcription factor GVG consists of the yeast Gal4 binding domain, the Herpes simplex VP16 activation domain, and the rat glucocorticoid receptor. Induction is based on the property of the glucocorticoid receptor domain to localize the constitutively expressed GVG protein to the cytoplasm in the absence of the inducer (Aoyama and Chua, 1997; Ouwerkerk et al., 2001).

After treatment with glucocorticoids such as dexamethasone, GVG is transferred to the nucleus and interacts with the GVG recognition site, 4UAS, cloned upstream of the targeting gene; it subsequently then activates transcription of the targeting gene (Ouwerkerk et al., 2001). The advantages of GVG-regulated gene expression in plant biotechnology include: (i) the regulatory molecules, i.e. glucocorticoid hormones, that have no naturally occurring analogues in plant cells; (ii) transgene expression is induced by the addition of glucocorticoid hormones at micromolar levels, and they have no deleterious effects on transformed tissues of some plant species at suitable concentrations, however, Kang et al. (1999) observed significant defects in transgenic Arabidopsis; and (iii) transcriptional control by the GVG transcription factor is very tight (Aoyama and Chua, 1997; Ouwerkerk et al., 2001).

Although GVG-regulated gene expression has been characterized extensively in dicotyledonous and monocotyledonous plants such as tobacco, Arabidopsis, and rice (Aoyama and Chua, 1997; McNellis et al., 1998; Ouwerkerk et al., 2001; Reynolds, 1999), it has not been tested in gymnosperms. Dexamethasone is widely used in GVGregulated gene expression in angiosperms, but it may not be the best candidate to regulate the GVG transcription factor in gymnosperms.

It was therefore investigated whether it was possible to regulate the level of transgene expression controlled by the GVG transcription factor in conifer cells more precisely, by using various glucocorticoid analogues. The effects of six readily available glucocorticoids (dexamethasone, hydrocortisone, 6-methylprednisolone, prednisone, prednisolone, triamcinolone) on GVG transcription factor-driven transgene expression in transformed Virginia pine cell cultures were tested.

Materials and Methods

Plasmid Constructs

After the pINDEX3 binary vector, provided by Dr PBF Ouwerkerk and Dr AH Meijer (Institute of Molecular Plant Sciences, Leiden University, Clusius Laboratory, The Netherlands), was isolated and purified, the plasmid pINDEX3 was digested by SpeI and XhoI (Promega). The m-gfp5-ER fragments (Haseloff et al., 1997; Stewart, 2001) were amplified from plasmid pBIN-m-gfp5-ER provided by Dr CN Stewart and Dr J Haseloff. The restriction enzyme SpeI and XhoI sites were introduced by PCR with Stratagene's Pfu DNA polymerase (Stratagene). For amplification of m-gfp5-ER fragments, the gfp forward primer 59-AAACATGATGAGCTTTAAAGACTC-39 and the reverse primer 59-CTTCATTGTTTGATCACCTTGCATCC- 39 (Sigma) were used. A total of 200 ng of plasmid DNA was used as a template in a 50 ll PCR reaction mix.

The PCR mixture consisted of 200 lM each of dATP, dCTP, dGTP, dTTP, 250 ng of each primer, 5 U Stratagene's Pfu DNA polymerase, 1.5 mM MgCl2, and 5 ll 103 buffer [500 mM KCl, 100 mM TRIS–HCl (pH 9.0 at 25 8C), 1% Triton X-100, 15 mM MgCl2]. The PCR conditions were 95 8C for 5 min followed by 29 cycles at 95 8C for 60 s, 57 8C for 40 s, and 72 8C for 90 s. Cycling was followed with a final incubation of 72 8C for 10 min.

PCR products were purified and digested by SpeI and XhoI (Promega). Ligation of the purified insert (30 ng) and vector (90 ng) was conducted in 20 ll with 2 U T4 DNA Ligase (Roche Applied Science) at 16 8C for 16 h.

One ll of ligation products was used to transform 20 ll NovaBlue competent cells (Novagen). Overnight cultures from single clones were used to isolate plasmid pINDEX3-m-gfp5-ER. The purified plasmid was introduced into Agrobacterium tumefaciens EHA105 competent cells by electroporation (Bio-Rad). Transformation of cultured cells and glucocorticoid hormone treatments

Cell cultures of Virginia pine were prepared as described in Tang et al. (2001a). Cell cultures were maintained by weekly subculture of a 2.5 ml cell suspension into 25 ml liquid culture medium [13TE Salt Base (Tang et al., 2001b), 30 g l_1 sucrose, 500 mg l_1 myo-inositol, 2 mg l_1 2,4-D, 0.5 mg l_1 BA, pH 5.8) and incubated at 25 8C, with orbital shaking at 150 rpm, in darkness. Cell cultures were transformed using Agrobacterium tumefaciens strain EHA105 carrying the binary vector plasmid pINDEX3-m-gfp5-ER, as described in Tang et al. (2001b). Agrobacterium, grown for 24 h to an optical density (OD600 nm=0.8–1.0) of bacteria in 3 ml of YEP broth (Sambrook et al., 1989), was centrifuged and resuspended in liquid medium. Cell-suspension cultures were infected with bacterium cultures for 15 min. The infected cell-suspension cultures were harvested with 42.5 lm filter papers, transferred to 125 ml Erlenmeyer flasks, and co-cultivation was carried out for 1–2 d.

Agrobacterium was removed from the cultures by washing the cell suspension cultures in 50 ml sterile tubes with 500 mg l_1 timentin (ticarcillin/clavulanic acid 3:0.1, v:v, SmithKline Beecham, Philadelphia, PA) solution for 3 min. The wash was repeated five times. Transgenic cell cultures were suspended in 25 ml liquid medium containing 5 mg l_1 hygromycin and 500 mg l_1 timetin and were incubated for 7 d at 25 8C, with orbital shaking at 150 rpm in darkness. Cells grown from these cultures were transferred to 25 ml fresh liquid medium containing 5 mg l_1 hygromycin. Further subculturing was performed weekly on media containing only hygromycin. To obtain large quantities of transformed cell cultures for further analysis, selected cell-suspension cultures were again introduced on a liquid proliferation medium.

After 6 weeks, the cultures were actively producing 50–70 mg of tissue l_1 each week, and they were then used to prepare DNA for PCR and Southern blot analysis. Stable transgenic cell lines confirmed by PCR and Southern blot analysis were used for systemic, inducible, gene-expression experiments. For treatments with glucocorticoid hormones, transgenic cell lines were grown in liquid plant medium for 3 d, then transferred into fresh liquid medium supplemented with the different glucocorticoid hormones, prednisolone, prednisone, 6-methylprednisolone, dexamethasone, triamcinolone, and hydrocortisone at different concentrations (0.1, 0.5, 1, 3, 5, and 10 mg l_1) for 24 h.

Cells were washed thee times, transferred to fresh liquid medium, and were observed under the fluorescence microscope at 2, 4, 6, 12, 18, 24, 36, 48, 72, 96, 108, 120, and 144 h after application of the inducer, following the protocol described by Scott et al. (1999).

At 24, 48, and 72 h after treatment of glucocorticoids, cell cultures were harvested for qualitative and quantitative fluorescent microscopy, and for northern blot analysis. At 21 d after treatment of glucocorticoids, cell cultures were harvested by centrifuging at 8000 rpm for 15 min for the measurement of fresh weight. Cell cultures were dried at 95 8C for 3 d for the measurements of dry weight (Tang, 2000). Glucocorticoids (Sigma) were stored as 100mMsolution in dimethyl sulphoxide (DMSO) at _20 8C. Experiments were replicated at least four times. Polymerase chain reaction and Southern blot analyses of transgenic cultures Genomic DNA was extracted from 500 mg cell cultures of control and putative transgenic cell lines, using a Genomic DNA Isolation Kit (Sigma) following the manufacturer's protocol.

The PCR reactions were performed with a PTC-100TM Programmable Thermal Controller (MJ Research) using DNA extracted from putative transformed and control cell cultures, respectively. The primers used for PCR reactions are the gfp forward primer (gfpf) 59-AAACATGATGAG CTTTAAAGACTC- 39 and the reverse primer (gfpr) 59- CTTCATTGTTATATCA CCTTGCAT-39. A total of 200 ng of genomic DNA was used as a template in a 50 ll PCR reaction mix. The PCR reaction was conducted as previously described (Tang et al., 2001b). PCR products were separated by electrophoresis on 1.0% agarose gels in 13 TAE buffer (Sambrook et al., 1989) and were detected by fluorescence under UV light (302 nm) after staining with 0.1% ethidium bromide. A molecular marker of 1 kb (Gibco-BRL, Gaithersburg, MD, USA) was used. Southern blot analysis was conducted as previously described (Tang et al., 2001b). RNA isolation and northern blot analysis Total RNA was isolated from 1.5 g transgenic cell cultures harvested though 42.5 lm filter papers and ground in liquid nitrogen using the RNeasy Plant Mini Kit (QIAGEN Inc.) following the manufacturer's protocol. Ten lg RNA was separated by agarose-gel electrophoresis. Electrophoresis and northern blotting of RNAs were performed as described by Tang and Tian (2003). Baked blots were prehybridized in 1 M NaCl, 1% SDS, 10% dextran sulphate, and 50 lg ml_ denatured herring sperm DNA at 64 8C, washed with 0.13 SSPE (13 SSPE is 180 mM NaCl, 10 mM NaH2PO4, 1 mM EDTA, pH 6.5), 0.5% SDS at 45 8C, and autoradiographed. Probes (816 pb fragment of m-gfp5-ER DNA) were labelled by Digoxigenin (DIG) (Roche Diagnostics Corporation). Equal loading of RNA samples was verified on the control of 25S rRNA. Qualitative and quantitative analysis of inducible GFP expression. To determine the effect of different concentrations of dexamethasone and different time of treatments on GFP expression, transformed cell cultures were observed at various times after treatments. GFP expression in transgenic cell cultures was observed with a stereo dissecting microscope equipped with a fluorescence module consisting of a 100 W mercury lamp and GFP Plus excitation and emission filters (Leica, Heerbrugg, Switzerland). This system (excitation filter 480 nm; dichoic mirror 505 nm LP; barrier filter 510 nm LP) permits visualization of GFP following excitation by blue light. For quantitative fluorescence determinations of m-gfp5-ER activity, samples were examined with a LSM 510 Laser Scanning Microscope (Carl Zeiss, Inc., Thornwood, NY, USA) using excitation with the 488 nm Argon laser line and detection of emitted light between 500 nm and 520 nm. The confocal images of m-gfp5-ER expression cells were created in the Expert Mode. Fluorescence intensities of different samples were calculated from confocal images with the Zeiss LSM Image Examiner software. The fluorescence level was quantified separately for the whole cell by circumscribing the respective area as a region of interest. Thirty to fifty cells were used for each sample.

Results

Transgenic Cell Lines

To generate transgenic cell lines, 476 cell cultures derived from single embryos were infected with an Agrobacterium tumefaciens strain containing pINDEX3-m-gfp5-ER carrying a modified hygromycin phosphotransferase gene, a chimaeric transcription factor GVG, and a modified GFP protein with an endoplasmic reticulum targeting sequence.

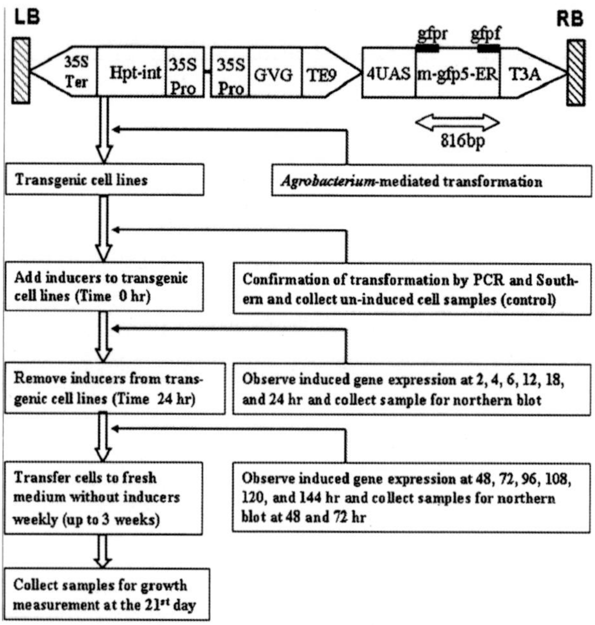

Figure: *A diagram of the experimental protocol including a linear plasmid map indicating the localization of the different genes [Hpt-int, hygromycin phosphotransferase gene with a intron from the castor bean Cat-1 gene; GVG, the chimaeric transcription factor consists of the yeast Gal4 binding domain (BD), the Herpes simplex VP16 activation domain (AD), and the rat glucocorticoid receptor (GR); m-gfp5-ER, a modified GFP protein with an endoplasmic*

Forty-four independent Virginia pine cell lines that were transformed with the pINDEX3-m-gfp5-ER plasmid (Ouwerkerk et al., 2001) and were resistant to hygromycin, were generated. After the T-DNA insert was confirmed by both PCR and Southern blot analyses (data not shown), six transgenic cell lines (P2, P5, P12, P16, P25, and P32) with 1–2 copies of the transgene were selected as candidates for inducible gene-expression experiments.

A summary of inducible GFP expression in all six transgenic cell lines was demonstrated in. After measurement of growth and cell survival rates at the end of the subculture period (7 d), one transgenic cell line (P32) with high growth and cell survival rates and containing one copy of the T-DNA insert, was selected for inducible gene expression. This transgenic cell line was transferred weekly into fresh proliferation medium for 10 weeks to produce more cells. No background GFP expression was observed in non-transformed control cell lines and all transgenic cell lines in the absence of inducer reticulum targeting sequence], promoters (4UAS, the promoter containing four copies of the GAL4 UAS and the _46 to +1 region of the 35S promoter; 35Spro, the cauliflower mosaic virus 35S promoter), and terminators [35Ter, the terminator of the CaMV 35S transcription unit; TE9, the poly(A) addition sequence of the pea ribulose biphosphate carboxylase small subunit rbcS-E9; T3A, the poly(A) addition sequence of the pea rbcS-3A], and T-DNA borders

(LB left border, and RB right border). Arrows indicate gene translation orientation. The probe used in northern blot analysis of transgenic cell lines is the PCR fragment of the m-gfp5-ER gene. Binding sites of PCR primers gfpr and gfpf are shown as black rectangles; their positions are indicated immediately above the m-gfp5-ER gene, and the predicted product size (816 bp) from gfpr and gfpf is shown as a double-headed arrow.

Inducible GFP Expression in Transgenic Cells

To assess the glucocorticoid, hormone-inducible, geneexpression system, the selected transgenic cell line was used to test inducible gene expression by six different glucocorticoid hormones (prednisolone, prednisone, 6-methylprednisolone, dexamethasone, triamcinolone, and hydrocortisone) at different concentrations (0.1, 0.5, 1, 3, 5, and 10 mg l_1). After the transgenic cell line was treated with glucocorticoid hormones for 24 h, cells were washed thee times and transferred to fresh liquid medium.

Transgenic cells were observed with the aid of a fluorescence microscope at 2, 4, 6, 12, 18, 24, 36, 48, 72, 96, 108, 120, and 144 h after the application of the inducer. Inducible gfp activity was observed at 4 h after treatment with dexamethasone and triamcinolone at concentrations of 1, 3, 5, and 10 mg l_1, and at 6 h after treatment of glucocorticoid hormones dexamethasone and triamcinolone at concentrations of 0.1 and 0.5 mg l_1.

Green fluorescence was observed at 6 h after treatment with prednisolone, prednisone, 6-methylprednisolone, and hydrocortisone at concentrations of 1, 3, 5, and 10 mg l_1, and at 12 h after treatment with prednisolone, prednisone, 6-methylprednisolone, and hydrocortisone at concentrations of 0.1 and PCR analysis of DNA isolated from transgenic cell lines. Lane M, 1 kb DNA molecular markers (Gibco-BRL); lane P, pINDEX3-m-gfp5-ER plasmid control; lane C, non-transgenic cells control; lanes 1–6, transgenic cell lines P2, P5, P12, P16, P25, and P32. The presence of the 816 bp band amplified by primers gfpr and gfpf from templates obtained from transgenic cell lines and absence of the 816 bp band amplified by primers gfpr and gfpf from templates obtained from non-transformed control cells indicate that the T-DNA is integrated into the pine genome.

GFP activity in all six transgenic cell lines 24, 48, and 72 h after treatment with 1 mg l_1 dexamethasone Data represent the mean 6SD. Values followed by different letters are significantly different (a=0.05) by ANOVA. 0.5 mg l_1. Transgenic cells produced green fluorescence upon blue light excitation 24 h after treatment with prednisolone, prednisone, 6-methylprednisolone, dexamethasone, triamcinolone, and hydrocortisone at all six concentrations tested. Inducible expression of m-gfp5-ER was visible in cultures incubated in all inducer concentrations.Asample of inducible m-gfp5-ER expression 24 h after treatment with prednisolone, prednisone, 6-methylprednisolone, dexamethasone, and triamcinolone at concentrations of 0.5, 1, 3, 5, and 10 mg l_1. images, 24 h after treatment with prednisolone, prednisone, 6-methylprednisolone, dexamethasone, and triamcinolone at concentrations of 0.1 mg l_1 and from hydrocortisone at different concentrations (0.1, 0.5, 1, 3, 5, and 10 mg l_1) were not included because of the dimness of the fluorescence. Inducible gene expression was also analysed by northern blot analysis of transgenic cell cultures treated with glucocorticoid hormones at a concentration of 1 mg l_1.

The results demonstrate that a high constitutive level of GVG gene expression was conferred by the CaMV35S promoter, and that m-gfp5-ER mRNA only appeared after glucocorticoid treatment (data not shown). Among samples taken 24, 48, and 72 h after application of 1 mg l_1 glucocorticoids, higher transcripts were observed from treated transgenic cell cultures at 48h. To evaluate the effects of glucocorticoid-inducible gene expression on cell growth and development, transgenic cell cultures were treated with concentrations of 0.1, 0.5, 1, 3, 5, and 10 mg l_1 for 24 h, and then subcultured weekly on liquid medium without inducer. Transgenic cell cultures were harvested at 21 d after glucocorticoid treatment for the measurement of fresh and dry weights. There were no significant differences in fresh and dry weight increases when transgenic cell cultures were treated with all six glucocorticoids at concentrations of 0, 0.1, 0.5, 1, 3, and 5 mg l_1. These results have demonstrated that cell growth and development were not inhibited by the addition of these compounds at suitable concentrations.

However, these results have demonstrated that fresh and dry weight increases of transgenic cell cultures were reduced by prednisolone, prednisone, 6-methylprednisolone, dexamethasone at concentration of 10 mg l_1, but growth was not inhibited by triamcinolone and hydrocortisone at a concentration of 10 mg l_1 (Tables 2, 3). The reduction in growth caused by the high levels of glucocorticoids was probably due to potentially toxic effects on cell cultures.

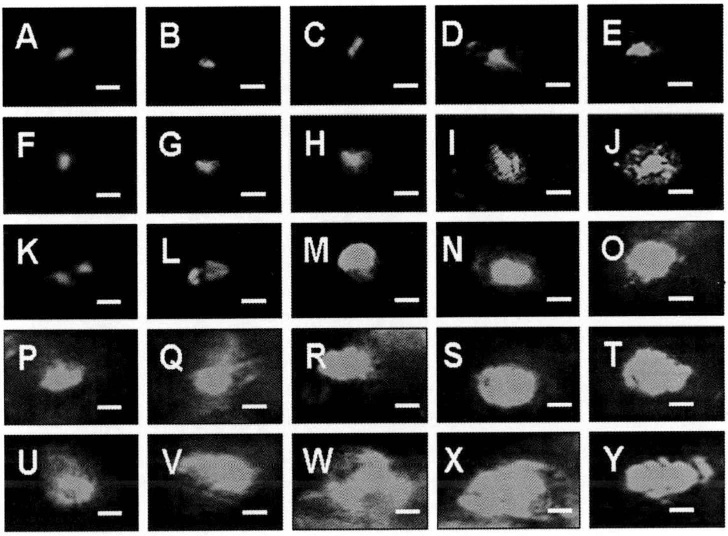

Figure: (A–Y) Fluorescence detection of GFP expression induced by prednisolone, prednisone, 6-methylprednisolone, dexamethasone, and triamcinolone, in a transgenic cell line. (A–E) GFP expression from P32 transgenic cells induced by different concentrations (0.5, 1, 3, 5, and 10 mg l_1) of prednisolone 24 h after treatment; (F–J) GFP expression from P32 transgenic cells induced by different concentrations (0.5, 1, 3, 5, and 10 mg l_1) of prednisone 24 h after treatment; (K–O) GFP expression from P32 transgenic cells induced by different concentrations (0.5, 1, 3, 5, and 10 mg l_1) of 6-methylprednisolone 24 h after treatment; (P–T) GFP expression from P32 transgenic cells induced by different concentrations (0.5, 1, 3, 5, and 10 mg l_1) of dexamethasone 24 h after treatment; (U–Y) GFP expression from P32 transgenic cells induced by different concentrations (0.5, 1, 3, 5, and 10 mg l_1) of triamcinolone 24 h after treatment. No GFP fluorescence was observed in transgenic cells without the addition of prednisolone, prednisone, 6-methylprednisolone, dexamethasone, and triamcinolone and in non-transformed control (bars=0.1 mm).

Regulated gene expression by glucocorticoids 1503 Quantitative analysis of inducible gfp expression To analyse glucocorticoid-inducible gene expression in further detail, m-gfp5-ER activities were determined quantitatively.

The effects of prednisolone, prednisone, 6-methylprednisolone, dexamethasone, triamcinolone, and hydrocortisone (at concentrations of 0.1, 0.5, 1, 3, 5, and 10 mg l_1) on m-gfp5-ER activities of transgenic cell cultures, were investigated.

Transgenic cell cultures were subcultured on liquid medium with different concentrations for 24 h, and green fluorescence of transgenic cells was determined quantitatively. Confocal images were taken at different times after treatment with dexamethasone Inducible gene expression turns off completely 144 h after the application of inducer.

Inducible m-gfp5-ER activities ranged between 63 and 80 (at 0.1 mg l_1), 66 and 92 (at 0.5 mg l_1), 79 and 127 (at 1 mg l_1), 103 and 190 (at 3 mg l_1), 105 and 192 (at 5 mg l_1), and 106 and 193 (at 10 mg l_1) units of relative fluorescence 24 h after treatment of prednisolone, prednisone, 6-methylprednisolone, dexamethasone, triamcinolone, and hydrocortisone.

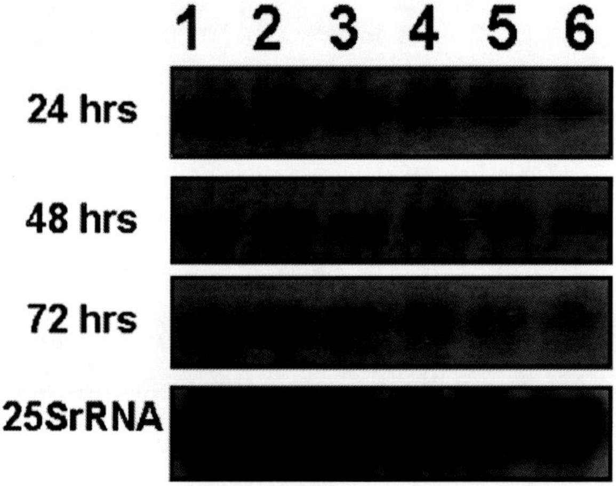

Figure: *Northern blot analysis of total RNA from p32 transgenic cell lines. RNA (15 lg) was extracted from transgenic cell lines treated with 1 mg l_1 prednisolone (line 1), prednisone (line 2), 6-methylprednisolone (line 3), dexamethasone (line 4), triamcinolone (line 5), and hydrocortisone (line 6) at 24 h, 48 h, and 72 h after treatment; they were hybridized (at 65 8C) with the 816 bp m-gfp5-ER probe corresponding to the m-gfp5-ER gene, which was labelled with Digoxigenin (DIG) (Roche Diagnostics Corporation, Roche Applied Science, Indianapolis, IN, USA). The integrity and the amount of RNA applied to each lane were verified by the control of 25SrRNA (lower panel).*

The highest level of green fluorescence was from transgenic cell cultures treated with3–10 mg l_1 triamcinolone for 24 h. In the absence of inducer, no m-gfp5-ER activity could be detected either qualitatively or quantitatively.

Glucocorticoid-inducible gene expression was also quantitatively determined in transgenic cell cultures treated with prednisolone, prednisone, 6-methylprednisolone, dexamethasone, triamcinolone, and hydrocortisone, for different intervals (24, 48, and 72 h) after treatment, respectively. No GFP fluorescence was detected in transgenic cell cultures that were not treated with glucocorticoids. However, background fluorescence, not from GFP expression but from the reflection of the light source, was observed.

The same amount of background fluorescence (20–28 arbitrary units in intensity; Table 1; was observed in both nontransgenic cells and transgenic cells before the application of the inducer. The differences in inducible gfp activity were observed among different glucocorticoids and at different times after treatment. GFP fluorescence was initially undetectable and increased from 0 h to 48 h after application of inducer.

In agreement with the northern blot analyses data, transgenic cell cultures from 24–48 h after treatment at 1 mg l_1 glucocorticoids showed a greater increase of fluorescence compared with the increase from 0–24 h and from 48–72 h. In general, high levels of green fluorescence were observed with transgenic cell cultures treated with dexamethasone and triamcinolone. Triamcinolone appeared to be the best inducer for the transgenic cell line investigated here. However, the dose and time required to saturate the system will be largely dependent on the growth stage and transgenic cell lines in different plant species used.

Discussion

Chemically inducible systems that activate or inactivate gene expression have many potential applications in plant biotechnology, particularly in the determination of gene function (Padidam, 2003). The precise timing and control of gene expression are important aspects of chemically inducible systems. Several systems-based chemical inducers have been developed and used. Chemicals that are used to regulate transgene expression include the antibiotic tetracycline (Gatz, 1997, 1999; Bohner et al., 1999), the steroids dexamethasone and estradiol (Schena et al., 1991; Aoyama and Chua, 1997), copper (McKenzie et al., 1998; Reynolds, 1999), ethanol (Caddick et al., 1998; Salter et al., 1998), ecdysone (Martinez et al., 1999), the inducer of pathogen-related proteins benzothiadiazol, herbicide safeners, and the insecticide methoxyfenozide (Padidam, 2003). Transgenic technology involving the introduction of one or more transgenes can turn desired traits on or off in animals and plants (Gatz, 1997; Gossen et al., 1995; Uknes et al., 1993). Although constitutive promoters are used to transcribe a gene of interest, a major limitation of constitutive promoters is that they cannot be used to investigate genes whose constant over- or under-expression has deleterious effects on the animals and plants where expression of a sense or anti-sense transgene in transformed cells may be toxic (Friedrich et al., 1996; Gatz and Lenk, 1998; Gatz, 1999; Picard et al., 1990; Reynolds, 1999).

Chemical-inducible systems for regulated gene expression offer a more general and flexible solution because chemical-inducible systems are quiescent in the presence or absence of inducers, and therefore will not inhibit physiological activities (Aoyama, 1999; Gatz et al., 1992; Thompson and Myatt, 1997). In addition, the use of an appropriate promoter to express the chemical-responsive transcription factor can further restrict the target transgene expression to specific organs, tissues, or even cell types (Faiss et al., 1996; Gossen et al., 1993). Chemical-inducible systems for regulated gene expression are extremely useful for basic biology research and biotechnology applications (Schena et al., 1991; Williams et al., 1992). In E.coli, an example of a promoter that provides such fidelity is the isopropyl b-D-thiogalactoside (IPTG) inducible lac promoter, which is routinely used whenever expression of the recombinant protein is going to interfere with growth (Beckwith and Zipser, 1970; Benhamou, 1996). Similarly, if a foreign gene product expressed in animals and plants is going to interfere with regeneration, growth or reproduction, an inducible promoter is required (Gatz, 1997). With such a tool, plants can be regenerated while the promoter is inactive. Analysis can then be performed after activating expression of the transgene (Zuo and Chua, 2000; Reynolds, 1999).

Functional genomics of 3-hydroxy-3-Methylglutaryl Coenzyme a Reductase Gene Family The symbiosis between leguminous plants and bacteria collectively named rhizobia leads to the formation of nitrogen-fixing root nodules. Depending on the host plant, nodules can be of determinate or indeterminate type, for which Lotus japonicus and Medicago truncatula have been selected as models, respectively. The first interaction between rhizobia and the host plant occurs at the root hair level in a restricted root zone that is competent for nodulation. Rhizobia attach to the root hair tip that curls and entraps the bacteria, which then enter the root hairs through the formation of infection threads. Infection threads progress through epidermal cells and reach the root cortex. Meanwhile, cortical cells dedifferentiate and start to divide leading to the formation of a nodule primordium. During the differentiation of an indeterminate nodule primordium, an apical nodule meristem is established (nodule zone I). Postmitotic cells exiting from the meristem continually become infected by rhizobia via budding of the infection threads into organelle-like structures called symbiosomes. Infected cells differentiate along several cell layers of the so-called nodule zone II until they reach their fully differentiated and nitrogen-fixing state in zone III. Initiation and development of nodules is mediated by signal exchanges between the host plant and its rhizobial partner. This molecular dialog controls the specificity of the interaction, nodule organogenesis, and the infection process. The earliest signals are flavonoid and isoflavonoid molecules produced by the host plant. Interaction of these plant signals with rhizobial NodD transcription factors activates the expression of rhizobial nodulation genes, which leads to the production of bacterial lipochitooligosaccharidic signals named Nod factors. Perception of Nod factors by the host plant induces many early events related to infection thread formation and primordium development (D'Haeze and Holsters, 2002).

In recent years, forward genetics and map-based cloning approaches have identified major components of the Nod factor signaling pathway in M. truncatula and L. japonicus (for review, see Stacey et al., 2006). Receptor-like kinases, such as Lj NFR1, Lj NFR5, or Mt NFP, with chitin binding LysM motifs in their extracellular domain are most likely the receptors for the chitin-like Nod factors since the corresponding mutants do not show any response to Nod factors (Madsen et al., 2003; Radutoiu et al., 2003; Arrighi et al., 2006; Smit et al., 2007). Immediately downstream of these Nod factor receptors, another receptor-like kinase known as NORK (also called Mt DMI2 or Lj SYMRK) is essential for transmitting the Nod factor signal (Endre et al., 2002; Stracke et al., 2002). Downstream of the Nod factor receptors and genetically unresolved from NORK, putative ion channels known as Mt DMI1, Lj POLLUX, and Lj CASTOR are required for the release of Ca from internal stores (Ané et al., 2004; Imaizumi-Anraku et al., 2005). The Ca fluxes take the form of spikes (periodic peaks and valleys of Ca concentrations) in the perinuclear and nuclear regions (Ehrhardt et al., 1996). Ca spikes are believed to be interpreted in the nucleus by a calcium/calmodulin-dependent protein kinase known as DMI3 (Lévy et al., 2004; Mitra et al., 2004; Tirichine et al., 2006). Finally, the activated calcium/calmodulin-dependent protein kinase stimulates NSP1, NSP2, and NIN transcription factors, which leads to changes in expression of symbiotic genes (Schauser et al., 1999; Kaló et al., 2005; Smit et al., 2005; Marsh et al., 2007).

This cell-autonomous or autocrine pathway operates in root hair epidermal cells that are directly receiving the Nod factor signal. However, Nod factors also trigger long-distance responses in the cortex leading to cell divisions (paracrine signaling), which are likely generated by secondary signals after Nod factor perception in the root hairs. Physiological data have implied cytokinins in the induction of nodule-specific gene expression and cortical cell divisions (Cooper and Long, 1994; Bauer et al., 1996; Fang and Hirsch, 1998; Mathesius et al., 2000). Using a cytokinin-responsive promoter controlling a marker gene, it was shown that the endogenous cytokinin levels increase in Rhizobium-stimulated root hairs and nodule primordia (Lohar et al., 2004). Genetic evidence for the role of cytokinin was provided recently, demonstrating that a His kinase cytokinin receptor is required and sufficient to induce cortical cell divisions and the establishment of a nodule primordium (Gonzalez-Rizzo et al., 2006; Murray et al., 2007; Tirichine et al., 2007). These findings suggest that Nod factors generate a localized production of cytokinins that could act as secondary signals for nodule primordium formation. However, it remains unclear how Nod factors stimulate cytokinin production.

Forward genetics approaches have been fruitful in the study of Nod factor signaling in legumes. Nevertheless, the nodulation process (especially the Nod factor signal transduction pathway) probably interacts with other pathways necessary for general cellular functions and plant development. Such interactions may be difficult to identify through genetic screens since mutations affecting such processes may be lethal or lead to severe pleiotropic phenotypes.

On the other hand, genes with redundant functions are rarely identified by such approaches. In this study, we aimed to identify novel elements of the Nod factor signal transduction pathway, focusing on the receptor-like kinase NORK (Endre et al., 2002; Stracke et al., 2002). NORK possesses an extracellular sensor domain with three Leu-rich repeats generally involved in protein–protein interactions and an NSL domain (NORK extracellular-sequence-like) with unknown function, a transmembrane domain, and an intracellular kinase domain. No ligands of NORK have been identified so far.

In response to purified Nod factors or rhizobia, NORK null mutants are unable to elicit calcium spikes, nodulin gene expression, or cortical cell divisions (Catoira et al., 2000; Wais et al., 2000; Esseling et al., 2004). Knocking down NORK expression by RNA interference (RNAi) in M. truncatula and Sesbania rostrata revealed that NORK is essential not only for epidermal functions but also for the release of bacteria from infection threads and symbiosome formation (Capoen et al., 2005; Limpens et al., 2005). In line with these observations, NORK expression was detected in the infection zone of M. truncatula nodules (Bersoult et al., 2005; Limpens et al., 2005). In this region, the NORK protein was localized to the plasma membrane and to the infection thread membrane (Limpens et al., 2005).

To gain insight into NORK function, we searched for proteins that interact with it. Using yeast two-hybrid (Y2H) screenings and immunoprecipitations, we identified 3-hydroxy-3-methylglutaryl CoA reductase 1 (Mt HMGR1) as a specific interactor of the NORK kinase domain. We show that this enzyme is crucial for nodule development. Our data couple the production of isoprenoid compounds (such as cytokinins and steroids) to Nod factor signaling and thus provide critical information toward understanding the function of NORK in symbiotic signaling and nodule development.

Results

The M. Truncatula NORK Protein Interacts with an HMGR

To identify interacting partners of NORK, Y2H screenings were performed with the extracellular 540 amino acids from the N terminus and the intracellular 383 amino acids from the C terminus domains of NORK using a Y2H cDNA library made of young nodules and roots of M. truncatula line R108 (Györgyey et al., 2000). Interactions were tested in repeated experiments under stringent selective conditions. Screenings with the extracellular domain of NORK did not result in the identification of interacting clones. By contrast, screenings with the intracellular domain (NORK-IC) revealed three independent positive clones. Each of them corresponded to the same HMGR gene that we named Mt HMGR1. These three clones were partial but overlapping and provided similar yeast growth under stringent selective conditions.

HMGRs are conserved enzymes of the mevalonate pathway and control the biosynthesis of diverse isoprenoid compounds in eukaryotes. The C terminus of HMGRs contains a catalytic domain, which allows a NADPH-dependent reduction of 3-hydroxy-3-methyl-glutaryl-CoA into mevalonic acid. This catalytic region was present in the three yeast clones that were identified. The longest clone (HMGR1ÄN), encoding 379 amino acids, was used to verify the NORK-IC–HMGR1 interaction. It was confirmed both in yeast by pairwise two-hybrid interactions and also by coimmunoprecipitations of the proteins from extracts of Arabidopsis thaliana protoplasts. In the latter case, the Mt HMGR1ÄN and NORK-IC sequences were cloned under the control of a cauliflower mosaic virus 35S promoter and tagged with different epitopes. An N-terminal 3HA tag was added to NORK-IC, while HMGR1ÄN was fused to the green fluorescent protein (GFP) and a c-Myc epitope (GFP-MYC tag) in the C terminus. HA-NORK-IC, HMGR1ÄN-GFP-MYC, and the GFP-MYC and HA-GFP control constructs were expressed transiently in Arabidopsis protoplasts. The presence of tagged proteins in the transfected protoplasts was verified by protein gel blot analysis. GFP was produced in high, HMGR1ÄN-GFP-MYC in medium, and HA-NORK-IC in lower amounts. To test the interactions, coexpressed or mixed protein extracts were immunoprecipitated with a mouse anti-HA antibody and separated by SDS-PAGE. The presence of c-Myc–tagged proteins was revealed with an anti-c-Myc antibody by protein gel blot analysis.

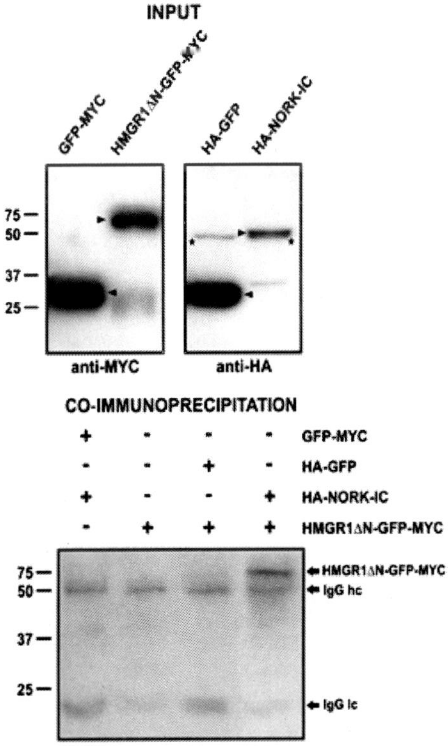

In all samples, the heavy and light chains of the anti-HA antibody used for immunoprecipitation were detectable as background signals due to their cross-reactivity with the secondary antibody. Nevertheless, the interaction between HA-NORK-IC and HMGR1ÄN-GFP-MYC was clear and specific, as no signal was observed between HA-NORK-IC and GFP-MYC or between HA-GFP and HMGR1ÄN-GFP-MYC. Consequently, we demonstrated in two independent experimental systems that NORK-IC interacts with the catalytic domain of HMGR1.

View larger version (35K):

In this window

In a new window

As a PowerPoint slide

Mt HMGR1ÄN Coimmunoprecipitates with Mt NORK-IC.

Top panel: Arrowheads indicate c-Myc–tagged (GFP-MYC and HMGR1ÄN-GFP-MYC) and HA-tagged (HA-GFP and HA-NORK-IC) proteins produced in Arabidopsis protoplasts by protein gel blot analysis using anti-MYC and anti-HA antibodies, respectively. In the HA-tagged protein extracts, asterisks indicate an aspecific band that runs slightly faster but was clearly different from HA-NORK-IC.

Bottom panel: Detection of c-Myc–tagged proteins in the anti-HA immunoprecipitates. Immunoglobulin G heavy chain (IgG hc) and light chain (IgG lc) of rat anti-HA antibody used for the immunoprecipitation are present in all samples and detected by weak cross-reaction with the sheep anti-mouse antibody. + and – indicate the presence or the absence, respectively, of given proteins in the reactions.

HMGRs Are Encoded by a Multigenic Family in M. truncatula

The number of genes encoding HMGRs varies significantly among plant species. The NORK-IC interacting partner isolated by Y2H derives from the R108 ecotype of M. truncatula. However, the ongoing genome-sequencing program uses the Jemalong A17 reference line. Homology searches in EST and genomic databases of M. truncatula allowed the identification of seven distinct homologs of Mt HMGR1 in Jemalong A17.

Five of them are full-length sequences, and two of them are partial ones. Since the genome sequence has not been completed, M. truncatula may contain even more HMGR genes. Although the catalytic domains are well conserved in the five complete HMGR sequences, significant variations in enzymatic function are possible due to the presence of nonhomologous amino acid replacements in the catalytic domains. Unlike mammalian HMGRs, which have eight N-terminal membrane-spanning domains, M. truncatula and other plant HMGRs only have two transmembrane domains.

These two domains and the linker between them are well conserved among Mt HMGR isoforms, but the N terminus and the linker between the second transmembrane domain and catalytic domain are highly variable. Based on homology scores and phylogenetic analysis, a single sequence of Jemalong A17 (TC106633) corresponded to Mt HMGR1 of R108. This Jemalong A17 sequence was used for subsequent studies. TM1 and TM2 are the hydrophobic transmembrane domains. Lines above the sequence alignments indicate HMGR1ÄN (broken line), HMGR1ÄNÄC1 (dotted line), and HMGR1ÄNÄC2 (full line). Asterisks mark positions that are variable in the NORK interaction domain among the five Mt HMGR proteins. NORK Interacts Specifically with the HMGR1 Isoform of the Multigenic HMGR Family Since the highly conserved catalytic region of HMGR1 interacted with NORK-IC, we tested whether the binding is specific for HMGR1 or occurs with other members of the Mt HMGR family. Interaction of NORK-IC with the catalytic domain of the five full-length Jemalong HMGR proteins was tested by pairwise Y2H assays. Despite the strong conservation of HMGR catalytic domains, the strength of interactions was remarkably different with the HMGR isoforms. Similarly to HMGR1 from R108, HMGR1 from Jemalong showed strong interaction with NORK-IC as yeast growth was confluent within 3 d, while no yeast growth was observed in the other pairwise interactions. After 6 d, however, yeast growth was also observed in the presence of HMGR2 (represented by EST contigs TC106634 and TC106637) and in a lesser extent in the presence of HMGR3 (TC106632 and TC106636), indicating that these Mt HMGR isoforms may bind weakly to NORK-IC.

By contrast, HMGR4 (TC106631) and HMGR5 (TC106635) did not show any interaction with NORK-IC. In our Y2H system, interaction of tested proteins also activated the expression of a lacZ reporter gene. Therefore, the strength of interactions was measured through the â-galactosidase activity of yeast colonies. Consistent with the yeast growth, the activity was high in the presence of HMGR1 and low in the case of HMGR2 or HMGR3. No â-galactosidase activity was measured in yeasts expressing HMGR4 or HMGR5. Thus, these results point to a specific interaction between Mt NORK-IC and HMGR1 in the Y2H system.

(A) Interaction of the NORK-IC with the different members of the Mt HMGR family in the Y2H system was evaluated by the growth (after 3 and 6 d) and â-galactosidase activity of yeasts.

(B) The different constructs of HMGRs and NORK or NORK homologs used for pairwise Y2H interactions. The size of the different derivates and the special motifs in the proteins are indicated. TM, transmembrane domain; SP, signal peptide; LRRs, Leu-rich repeats; AA, amino acids. The asterisk indicates the mutation in NORK-R38.

(C) Results of the pairwise Y2H interactions. ++ indicates yeast growth at 3 d (strong interaction), + indicates yeast growth at 6 d (weak interaction), and – indicates no yeast growth (no interaction).

The N-Terminal Half of the Mt HMGR1 Catalytic Region Is Sufficient and Specific for Binding to Mt NORK and Its Putative Orthologs from Other Legumes. To delimit the protein regions responsible for the interaction between NORK and HMGR1, deletion derivatives of both proteins were created and tested in the Y2H system in pairwise interactions. The 3' deletions in HMGR1ÄNÄC1, reducing the 379–amino acid region from the C-terminal part to 280 or 181 amino acids, had no effect on binding, demonstrating that the NORK binding site resides within the N-terminal part of the catalytic domain.

The NORK-IC contains 383 amino acids from the C terminus of NORK. Separating this region into two overlapping halves, one containing 228 N-terminal amino acids and the other 266 amino acids from the C terminus, abolished the interaction with HMGR1. Thus, either a longer or the entire intracellular domain of NORK is needed for this interaction or for correct folding of the binding domain. To test whether the binding requires the NORK kinase activity, the R38 Nod mutant allele of NORK (dmi2-4) was tested by Y2H. This mutation, a Gly-to-Glu amino acid replacement (Endre et al., 2002) converted the protein to an inactive kinase as in vitro autophosphorylation activity of the wild-type NORK kinase domain was abolished in the R38 mutant (data not shown).

Using this clone, no interaction was detected with HMGR1ÄN, indicating that the NORK kinase activity is probably necessary for the interaction. This could either be required for substrate phosphorylation or self-phosphorylation of the NORK protein. Subsequently, we tested whether other kinases can substitute NORK in the interaction with HMGR1ÄN.

We chose three M. truncatula kinases (TC102886, TC103254, and TC108366) that showed the highest homologies with NORK in their kinase region at the current status of available genomic sequence. In pairwise combinations with HMGR1ÄN, the kinase domain of these proteins resulted in no yeast growth, indicating that HMGR1ÄN binds specifically to NORK-IC.

This raised the possibility that Mt HMGR1 may also bind putative orthologs of NORK from other legumes. Therefore, we cloned the intracellular regions (383 amino acids), equivalent to that of Mt NORK-IC, from L. japonicus (Lj SYMRK-IC) and S. rostrata (Sr SYMRK-IC). Although weaker than Mt NORK, both putative NORK orthologs interacted with Mt HMGR1ÄN in the Y2H system, providing support for the notion that the NORK/SYMRK–HMGR1 interaction is conserved among legumes.

Expression and Subcellular Localization of Mt HMGR1

The possible interaction of NORK and HMGR1 in vivo presumes their overlapping expression patterns during nodulation. Previous works demonstrated that the Mt NORK/DMI2 gene is expressed in the root, in the nodulation competent zone, in nodule primordia, and in nitrogen fixing nodules in a few cell layers of zone II adjacent to the meristem (Bersoult et al., 2005; Limpens et al., 2005; our unpublished data). At the subcellular level, NORK:GFP (DMI2:GFP) proteins have been localized to the plasma membrane and to the infection thread membrane (Limpens et al., 2005).

To obtain a global view on the HMGR genes, their expression was first tested in different M. truncatula organs by RT-PCR experiments. All five HMGR genes were expressed in roots and nodules.

Mt HMGR3, HMGR4, and HMGR5 were also transcribed in the aerial organs, such as leaves, stems, and flowers at varying levels. Like NORK, expression of Mt HMGR1 and HMGR2 was restricted to the root system, where HMGR1 transcript levels were more abundant in the root than in the nodules. This prompted us to test and compare expression patterns of HMGR1 and NORK during nodulation.

NORK, similarly to the constitutive Mtc27 gene, was expressed at a constant level in roots, Nod factor-treated roots, and nodule primordia as well as in young and mature nitrogen fixing nodules. By contrast, expression of HMGR1 increased in Nod factor–treated roots and in the nodule primordium and decayed in the nitrogen fixing nodules. NORK is expressed in the root hairs where Nod factor perception occurs and signaling events begin. Therefore, we studied expression of HMGR1 compared with NORK in isolated root hairs and in roots devoid of root hairs.

Both genes were expressed in root hairs and in roots lacking root hairs, in line with the postulated function of HMGR1 in the NORK signaling pathway. In situ hybridizations on root sections in the nodulation competent root zone confirmed colocalization of HMGR1 and NORK transcripts in the epidermal cell layer where Nod factor signaling takes place.

(A) Expression pattern of NORK and the HMGR genes in different plant organs by RT-PCR analysis. The constitutively expressed Mtc27 was used as control for RNA loading in (A) to (C). R, roots; N, nodules; L, leaves; S, stems; F, flowers.

(B) Expression pattern of NORK and HMGR1 in roots, Nod factor–treated roots (R+NF), nodule primordia (NP), young nitrogen fixing nodules (YNFN), and mature nitrogen fixing nodules (MNFN).

(C) Expression of NORK and HMGR1 in root hairs (RH) and roots devoid of root hairs (R-RH).

(D) Localization of NORK (NORK-AS) and HMGR1 (HMGR1-AS) transcripts by in situ hybridization in the nodulation-competent root zone. Dark color (arrows) corresponds to the hybridization signal, which is absent in the control section hybridized with the sense HMGR1 (HMGR1-S) probe. Bars = 50 μm.

(E) Subcellular localization of the YFP control and MtHMGR1-YFP in Arabidopsis protoplasts. DIC, differential interference contrast. Bars = 10 μm. To determine the subcellular localization of HMGR1, the protein was fused to yellow fluorescent protein (YFP) and expressed from the cauliflower mosaic virus 35S promoter in Arabidopsis protoplasts. The HMGR1-YFP signals were found predominantly on vesicle-like structures in this system.

This localization is in agreement with the immunolocalization of At HMGR1 in Arabidopsis cells where the protein was present on uncharacterized vesicles that most probably originate from the endoplasmic reticulum (ER) (Leivar et al., 2005). Pharmacological Inhibition of HMGR Activity Reduces the Nodulation Efficiency. Lovastatin (mevinolin) is a potent and widely used inhibitor of HMGR enzymatic activity (Alberts et al., 1980). To confirm the role of HMGR in nodulation, the effect of lovastatin treatment on M. truncatula nodulation was investigated. To select a maximal lovastatin concentration with minimal nonspecific effect on plant development, the growth capacity of the main root and formation of lateral roots were tested at different concentrations of lovastatin (0 to 5 μM). Lovastatin up to 0.5 μM concentrations had a negligible-to-weak effect on primary and lateral root growth.

Lovastatin at 1 μM strongly modified root architecture, reducing primary root growth and stimulating lateral root formation.

Lovastatin at 5 μM strongly inhibited the primary root growth, and although it induced the formation of numerous lateral root primordia, these primordia were arrested and never developed into lateral roots.

(A) Primary root growth at different concentrations of lovastatin. Small letters (a to d) in (A), (B), and (D) represent significantly distinct categories by Student's t test at 95% confidence.

(B) Lateral root formation at different concentrations of lovastatin. At 5 μM lovastatin, only lateral root primordia were formed that did not develop further.

(C) Growth of S. meliloti is unaffected by lovastatin.

(D) Nodulation efficiency in the presence of different lovastatin concentrations in three independent experiments.

Sinorhizobium meliloti (Sm1021), the symbiotic partner of M. truncatula, has no HMGR gene; therefore, it was unlikely that lovastatin affects its growth. Nevertheless, to exclude the formal possibility of a nonspecific effect, the growth rate of S. meliloti was measured and was found to be unaffected by lovastatin up to 5 μM.

Nodulation of M. truncatula by S. meliloti (Sm1021) was performed in the absence and presence of lovastatin at 0.5, 1, and 5 μM concentrations. As expected, the toxic 5 μM concentration of lovastatin inhibited root growth and nodulation. Three independent experiments showed that the presence of lovastatin at 0.5 or 1 μM significantly reduced the nodule numbers. Since 0.5 μM lovastatin did not significantly affect the root growth, this negative effect on nodulation indicates that HMGR activity is indeed required for nodule development.

RNAi of Mt HMGR1 Inhibits Nodulation

Since lovastatin is a general inhibitor of HMGR enzymes, our pharmacological studies did not specify which isoform is required for nodulation. Given the specific interaction observed between NORK and HMGR1 in the Y2H system, we studied the involvement of HMGR1 in nodulation by RNAi in transgenic hairy roots of M. truncatula. We cloned a 226-bp 5' untranslated region of Mt HMGR1 in a pK7GWIWG2(II)-derived binary plasmid vector carrying the DsRED1 fluorescent marker resulting in the pRNAiMtHMGR1 construct. Similarly, pRNAiMtHMGR4 was created by cloning a 158-bp 5' untranslated region of the noninteracting HMGR4 gene.

These plasmids, as well as the pRedRoot:GFP control plasmid (Limpens et al., 2004; Riely et al., 2007), were introduced into Agrobacterium rhizogenes MSU440. By the transfer of the A. rhizogenes Ri T-DNA, these strains provoked the formation of transgenic hairy roots on M. truncatula seedlings. The DsRED1 fluorescent marker indicated the cotransformed roots. In two sets of experiments, >120 wild-type (Jemalong A17) plants were transformed with each construct. Hairy roots were scored for nodule formation 3 weeks after the inoculation with S. meliloti. In the first set of experiments, seven to eight nodules were formed in average on the DsRED1 and the DsRED1 pRedRoot:GFP and pRNAiMtHMGR4 hairy roots.

By contrast, no nodules appeared on the cotransformed transgenic roots expressing pRNAiMtHMGR1, while on the same plant, roots that were not cotransformed with pRNAiMtHMGR1 developed a comparable number of nodules as the controls. In the pRNAiMtHMGR1 roots the expression of HMGR1 was reduced on average to 40% of the wild type level. Due to the high degree of similarity of the Mt HMGR genes, we have also tested the specificity of this RNAi strategy for Mt HMGR1. The Mt HMGR1 sequence used for RNAi was the most similar to Mt HMGR2; therefore, we measured the expression level of this gene in the pRNAiMtHMGR1 background and in the DsRED1 roots. Since the expression level of HMGR2 was not affected by pRNAiMtHMGR1, we could conclude that the RNAi silencing was specific for

Mt HMGR1 and that HMGR1 function is necessary for nodulation. In this set of experiments, the complete absence of nodulation was surprising, as one would expect with RNAi also milder effects on nodulation. By changing light intensity and humidity, we optimized the nodulation on hairy roots. Nodulation rates on wild-type or DsRED1 roots raised from 8.1 ± 2.9 nodules per root to 12.8 ± 2.3. Under these favorable conditions, a few nodules appeared also on the pRNAiMtHMGR1 roots (4 ± 1). However, these nodules only reached the primordium state. On very rare occasions, infection threads developed. They were usually arrested early in the root hairs or in the outer cortical cells and never lead to formation of nitrogen fixing nodules. These results thus confirmed an essential role for HMGR1 in nodulation and more particularly in the infection process that was consistent with the specific NORK–HMGR1 interaction.

(A) and (B) Transgenic M. truncatula roots expressing pRNAiMtHMGR1.

(C) and (D) Transgenic M. truncatula roots expressing pRNAiMtHMGR4.

Roots were observed under bright field in (A) and (C) and fluorescence microscope in (B) and (D) to identify hairy roots that were DsRED1 positive (DsRED1) and distinguish them from the DsRED1-negative (DsRED1; non-cotransformed) roots. Arrows indicate nodules. Bars = 0.5 cm.

(E) Nodule number at 3 weeks after inoculation on DsRED1 versus DsRED1 roots of pRNAiMtHMGR1, pRNAiMtHMGR4, or pRedRoot:GFP transformation. Nodulation was similar in all plants except the DsRED1 pRNAiMtHMGR1 roots, where no nodules developed.

(F) Relative expression levels of HMGR1 and its closest homolog HMGR2 measured by quantitative RT-PCR in DsRED1 roots expressing pRNAiMtHMGR1 compared with DsRED1 ones.

(G) Optimized hairy root nodulation allows nodule initiation on DsRED1 pRNAiMtHMGR1 roots but in reduced number and with no or aborted infection. Error bars in (E) to (G) represent SD.

(H) Nodules on DsRED1 pRNAiMtHMGR1 roots were mainly uninfected and halted in the primordium state.

(I) Occasional aborted surface infections marked by the blue color (arrow) of S. meliloti lacZ activity in a DsRED1 pRNAiMtHMGR1 nodule.

Discussion

The NORK–HMGR1 Interaction Is Specific and Conserved in Legumes Our work demonstrates a specific interaction between Mt NORK and HMGR1 protein. The HMGR1 binding site in NORK resides within the intracellular kinase domain and probably requires the active kinase domain since a putative inactive kinase mutant, NORK-R38 (dmi2-4) (Endre et al., 2002), did not bind HMGR1 in the Y2H system. Because the M. truncatula dmi2-4 mutant is unable to form nodules, the inability of NORK-R38 to interact with HMGR1 identifies HMGR1 as a potential downstream element of NORK signaling. The kinase region of Mt NORK could be substituted by putative NORK orthologs from other legumes, such as Sr SymRK or Lj SYMRK, which revealed the conservation of this interaction and supported further the role of HMGR1 in nodulation. Consistent with this hypothesis, kinase domains of NORK homologous, but nonsymbiotic receptor-like kinases, from M. truncatula did not interact with HMGR1. Thus, these data collectively show a specific interaction of the NORK kinase domain with HMGR1 and the conservation of this interaction in other legumes.

In M. truncatula, the HMGR family is composed of at least seven members. A strong and specific interaction of NORK was detected only with HMGR1. The NORK binding domain within HMGR1 was delimited to the first half of the catalytic region. Interestingly, this region is highly conserved among other members of the Mt HMGR family. The delimited 181–amino acid region of HMGR1 required for NORK binding differs only in three amino acids from the weakly interacting HMGR2, two of which are homologous replacements (Ile-264 to Val and Ala-287 to Ser), while Leu-220 to Asn is a nonhomologous one and most likely responsible for the drastically reduced binding ability of HMGR2 to NORK.

HMGR3 interacting even weaker with NORK differs from HMGR1 in seven amino acids, corresponding to three nonhomologous exchanges, Glu-180/Ser, Ser-303/Asn, Thr-315/Ile, and to four homologous ones, Arg-177/Lys, Ile-264/Val, Ala-287/Ser, and Thr-343/Ser. In noninteracting HMGR4, there are three nonhomologous replacements, Gln-283 to Leu, Ser-303 to Asn, and Met-308 to Ile, and three homologous ones, Ile-264/Val, Ala-287/Ser, and Leu-288/Ile. The same Gln-283/Leu and Met-308/Ile replacements are also present in noninteracting HMGR5, which possesses several other sequence differences as well. These Gln and Met residues are unaffected in the weakly interacting HMGR2 and HMGR3, suggesting that Gln-283 and Met-308 in HMGR1 are essential for interaction with NORK, while Lys-220 is crucial for the efficient binding.

Receptor-like kinases are the predominant class of cell surface receptors in plants. Their biological functions have been assigned in a wide range of processes, including development and interactions with the environment (Becraft, 2002; Morris and Walker, 2003). Nevertheless, the signaling cascades activated by plant receptor-like kinases remain poorly or partially defined. We expected to obtain a signal transduction protein as an interacting partner for the NORK receptor-like kinase. Finding instead the enzyme HMGR as partner was unforeseen. However, this interaction can be meaningful for different reasons as HMGR is highly regulated, for example, by phosphorylation by kinases or by specific subcellular compartmentalization (Stermer et al., 1994; Chappell, 1995; Leivar et al., 2005).

As only the active NORK kinase showed interaction with HMGR1 in the Y2H assay, kinase activity might be required for autophosphorylation of NORK or phosphorylation of HMGR1, which may lead to inactivation of the enzyme (Goldstein and Brown, 1990). While NORK kinase activity resulted in autophosphorylation, no phosphorylation of HMGR1 was detected in in vitro kinase assays (data not shown). Thus, binding of NORK to HMGR1 might keep HMGR1 in an active unphosphorylated form that could be critical for nodulation. On the other hand, autophosphorylation of NORK might provoke a conformation change required for the binding of HMGR1.Thus, NORK may function as a scaffold protein and attract the HMGR1 enzyme to a particular subcellular localization at the plasma membrane or infection thread membrane levels.

HMGR1 Activity Is Required for Nodulation in M. truncatula

The requirement of HMGR1 for nodulation is supported by pharmacological inhibition of HMGR activity during nodulation. Lovastatin is one of the most frequently used inhibitors of HMGRs. We showed that lovastatin at low concentrations, without visible effects on growth and development of M. truncatula, significantly reduced nodulation, confirming that nodule development indeed depends on an HMGR activity. The need for HMGR1 was further supported by RNAi experiments, which showed that an <“60% reduction in the HMGR1 transcript levels in M. truncatula roots strongly decreased nodule formation. These data collectively support a key function in nodulation for HMGR1 in a NORK-dependent manner. NORK is also required for mycorrhizal symbiosis, but there is little evidence for the involvement of HMGR1 in mycorrhizal development. One indication of a possible role in mycorrhization is that Mt HMGR1 (represented by EST AW584544) is listed among genes that were transiently upregulated in the initial stages of the mycorrhizal symbiosis (Liu et al., 2003).

Where Does NORK Interact with HMGR1?

Both NORK and HMGR1 are expressed in the root and during the early stages of nodule development. HMGR1 transcripts, like NORK transcripts, are present in the epidermal cells of the nodulation-competent root zone. These overlapping expression patterns support the notion that NORK and HMGR1 function in the same cells. NORK-GFP is plasma membrane bound (Limpens et al., 2005). The topology of NORK, similar to other receptor like kinases, predicts intracellular localization of the kinase domain. Bacterial HMGRs are localized in the cytoplasm, whereas mammalian HMGRs possess eight transmembrane domains and are ER bound. Plant HMGRs have two transmembrane domains, and the different HMGR isoforms are believed to localize to discrete and distinct domains of the ER or ER-derived vesicles (McCaskill and Croteau, 1998; Leivar et al., 2005). The topological model of HMGR suggests that the short stretch between the transmembrane domains is in the lumen, while the N terminus and the C-terminal catalytic domain are exposed to the cytosol (McCaskill and Croteau, 1998; Leivar et al., 2005).

The Mt HMGR1-YFP fusion protein did not localize to known organelles in Arabidopsis protoplasts; rather, it was found on vesicle-like structures. At HMGR1 was found on similar structures by imaging of a GFP fusion protein and immunolocalization of endogenous At HMGR1 (Leivar et al., 2005). The nature of the HMGR vesicles is undefined, but these subcellular compartments most likely originate in the ER (Leivar et al., 2005). Thus, the localization of Mt HMGR1 could be bona fide and not an artifact of the 35S promoter overexpression of the fusion protein.

If NORK is plasma membrane associated and HMGR1 is mainly on vesicles, how do they interact? Although they are anchored in different membranes, the interacting domains of both proteins are cytosolic and could be brought together by intracellular movement of the HMGR vesicles. Mutual recognition of the binding partners might also trigger controlled fusion of the plasma membrane with the vesicles, creating specific microdomains or compartments. Such membrane fusions are fundamental in eukaryotic cells and crucial for the transfer of proteins and lipids between different compartments and for exo- and endocytosis (Battey et al., 1999).

Thousands of isoprenoid compounds (derived from the common five-carbon building unit isopentenyl diphosphate and its isomer dimethylallyl diphosphate) have been identified with essential functions in many aspects of life. Cholesterol is probably the best known and most studied derivative of the isoprenoid pathway. Plants produce an exceptionally high variety of isoprenoids and isoprenoid derivatives, many of them playing key roles in plant growth and development, photosynthesis, or resistance to pests. In plants, as in bacteria, two pathways for isoprenoid biosynthesis exist (Rodríguez-Concepción and Boronat, 2002).

The mevalonic acid (MVA) pathway is cytosolic, and HMGR is considered to be the rate-limiting step. The methylerythritol 4-phosphate pathway is active in plastids. The methylerythritol 4-phosphate pathway assures synthesis of gibberellins, abscisic acid, carotenoids, and chlorophyll. Important metabolites derived from the MVA pathway, and thus under the control of HMGR, are cytokinins, phytosteroids, including brassinolins and the farnesyl, or geranylgeranyl moieties of the isoprenyl lipid protein modifications.

(A) The steps of the cytoplasmic isoprenoid synthesis with indication of the relevant intermediates and important end products (in italics and underlined). The enzymatic reaction catalyzed by HMGR (boxed) is shown as well as inhibition of HMGR by lovastatin.

(B) Model for potential implications of HMGR1 via NORK in nodule development.

In contrast with animals, plants have multiple HMGR isoforms. Their differential expression (Choi et al., 1992; Enjuto et al., 1995), subcellular localization (McCaskill and Croteau, 1998; Leivar et al., 2005), and mutant phenotypes (Suzuki et al., 2004) indicate that the distinct isoforms have metabolic specialization and, thus, that they may be involved in different branches of the mevalonate pathway and the production of different end products. This is likely achieved by metabolic channeling, meaning that cooperating enzymes are organized into macromolecular complexes in which biosynthetic intermediates are transferred between catalytic sites without diffusion into the bulk phase of the cell (Chappell, 1995). NORK interacts with HMGR1 and only weakly, or not at all, with other HMGRs. This indicates that the Nod factor signaling pathway activates or recruits a particular isoprenoid pathway. What kind of isoprenoid metabolites can be regulated by NORK-HMGR1 and involved in nodulation? Based on the metabolites of the MVA pathway and the physiology of nodule formation, we propose four alternative hypotheses.

Cytokinin is produced by isopentenyl transferases from adenosine phosphate and dimethylallyl pyrophosphate, a product from the mevalonate pathway. Therefore, cytokinin synthesis can be suppressed by lovastatin in cultured Nicotiana tabacum BY-2 cells (Crowell and Salaz, 1992). Cytokinin is known to play a major role in nodule initiation. Cytokinin likely is produced in Nod factor–stimulated root hairs and may serve as a secondary, mobile signal to the root cortical cells. Perception of cytokinin stimulates cell division in the cortex and formation of the nodule primordium (Gonzalez-Rizzo et al., 2006; Murray et al., 2007; Tirichine et al., 2007). Thus, it is possible that following Nod factor stimulation, the recruitment of HMGR1 by NORK in the root epidermis leads to the production of cytokinins for paracrine signaling and the stimulation of cortical cell divisions. If HMGR1 is the isoform that promotes cytokinin synthesis, then downregulation of HMGR1 would be expected to block cell division and to inhibit nodule formation, which was indeed the case in the HMGR1 RNAi plants. On the other hand, analysis of M. truncatula DMI3 gain-of-function mutants suggests that DMI3 activity is sufficient to initiate nodule development in a wild-type background. To our knowledge, it has not been shown that DMI3 gain-of-function mutants are able to induce nodulation in the absence of DMI2 or HMGR1. Although the requirement of both DMI3 and NORK-HMGR1 activities for cytokinin production seems unexpected, this hypothesis cannot be excluded.

Besides being part of the early signaling pathway operative in root hairs, NORK is also involved in later stages of nodule formation and, more precisely, in controlling the endocytotic uptake of rhizobia from infection threads into host cells (Capoen et al., 2005; Limpens et al., 2005). Endocytosis in plant cells is still poorly understood (Battey et al., 1999). The relationships between the endocytotic uptake of rhizobia and other endocytotic processes are unknown. In mammalian cells, where endocytosis is far better described, this process requires steroids and cholesterol in specialized membrane microdomains named lipid rafts (Heese-Peck et al., 2002; Thomsen et al., 2002; Pichler and Riezman, 2004). Such lipid rafts play an essential role in endocytosis but also in signaling processes and cytoskeleton organization (Funatsu et al., 2000; Brdickova et al., 2001).

The presence of lipid rafts in the plasma membrane of M. truncatula has been demonstrated recently, and these structures were found to be associated with a complete redox system at the plasma membrane level (Lefebvre et al., 2007). Mammalian pathogens, including Brucella species, which are closely related to rhizobia and whose uptake and intracellular life have common features with rhizobia in legume nodules (Batut et al., 2004), often hijack the cholesterol-dependent endocytotic pathway for their uptake into host cells (Gatfield and Pieters, 2000; Shin et al., 2000; Mañes et al., 2003; Batut et al., 2004). Thus, a second possible function of NORK-HMGR1 could be to direct a metabolic complex for steroid synthesis in the membrane of infection threads and thereby promoting locally the production and accumulation of steroid compounds that then mediates endocytotic uptake of the rhizobia from the infection threads. However, the significant decrease in nodule number of RNAi roots suggests a role for NORK-HMGR1 at an earlier stage than rhizobial release from infection threads. Possibly the NORK-HMGR complex is active at the stage of root hair infection, when rhizobia, entrapped in a root hair curl, induce an invagination of the plasma membrane to initiate infection thread growth.

In animals, inactivation of HMGR has profound effects on development and cell differentiation because of its effect on protein isoprenylation. Loss of HMGR activity, by mutation or pharmacologically, leads to the absence of farnesyl or geranylgeranyl posttranslational modifications and consequently mislocalization of important signaling proteins, such as small G-proteins involved in development. These lipid modifications are needed for membrane anchoring of these proteins and thus for their proper subcellular localization (Van Doren et al., 1998; Santos and Lehmann, 2004; Thorpe et al., 2004; Yi et al., 2006; D'Amico et al., 2007). Protein farnesylation and geranylgeranylation is also conserved in plants and is known to affect the subcellular localization and functioning of signaling proteins, such as ROP/RAC small G-proteins, transcription factors, cell cycle regulators, and others (Yalovsky et al., 1999, 2000; Crowell, 2000; Lavy et al., 2002; Galichet and Gruissem, 2003, 2006).

NORK could act as a scaffold attracting HMGR and other enzymes of the protein isoprenylation pathway to engage signaling proteins locally in a Nod factor signaling complex.

Finally, the production of isoprenoid compounds could assure the signaling between the plasma membrane–localized NORK and the nuclear-localized DMI1 protein (Riely et al., 2007). Based on homologies with bacterial ion channels (such as MthK), DMI1 is predicted to be a nuclear ion channel, and its activity may be regulated by calcium, NAD, or other unknown ligands (Ané et al., 2004; Riely et al., 2007). The presence of a complete redox system in M. truncatula lipid rafts (Lefebvre et al., 2007) suggests that redox signals and NAD production could be the missing link between NORK and DMI1 and more generally between the site of Nod factor perception and nuclear calcium spikes. It is also possible that DMI1 ligands may derive from the mevalonate pathway, which would provide a direct connection between DMI1 and HMGR activities. To elucidate the mode of action of HMGR1, more information is needed about the different branches of the mevalonate pathway and the end products produced by the different HMGR isoforms. In addition, the step in nodule formation and Nod factor signaling that is affected by HMGR1 must be defined more precisely.

Methods

Nodulation Assays

Medicago truncatula ecotype Jemalong (line A17) and ecotype R108 were used for nodulation assays. Seeds were immersed for 4 min in pure H_2SO_4 for scarification and sterilized with 8% (w/v) calcium hypochlorite solution for 20 min. Seeds were then placed for 3 d at 4°C in the dark and germinated overnight at room temperature in the dark. Freshly germinated seedlings were transferred to square plates containing BNM nodulation growth medium (Ehrhardt et al., 1992). Eight seedlings were placed per plate and the lower part of the plate was wrapped in aluminum foil to keep the roots in the dark. The plates were placed vertically in a growth room at 24°C in 16-h-light/8-h-dark cycle conditions. After 2 d of growth, the plants were transferred to fresh BNM plates containing lovastatin at the final concentrations as indicated. The 12.5 mM lovastatin (Sigma-Aldrich) stock solution was prepared freshly in ethanolic NaOH (15% [v/v] ethanol and 0.25% [w/v] NaOH) and incubated at 60°C for 1 h for hydrolyzing the lactone ring. After 1 d of growth on lovastatin-containing plates, the roots of the plants were inoculated with Sinorhizobium meliloti Sm1021 that was grown overnight in TA liquid medium (1% tryptone, 0.1% yeast extract, 0.5% NaCl, 1 mM $MgSO_4$, and 1 mM $CaCl_2$) at 30°C with 150 rpm up to OD_{600} H" 1.0 and then washed twice with distilled water and resuspended in water at OD_{600} H" 0.05. The number of lateral roots and the number of nodules were counted at different times as indicated. For measurement of the length of the main root, the growth path of the root was indicated on the Petri dish with a pencil and scanned and measured with Adobe Photoshop 7.0 tools.

Y2H Screening and Pairwise Assays

For the Y2H screenings and pairwise interactions, bait constructs were cloned in the vector pBD-GAL4 (Stratagene). The bait sequences for the Mt NORK genes and homologous kinases were obtained by PCR amplification of root or nodule cDNA using oligos with EcoRI and XhoI linkers for cloning in pBD-GAL4. Truncated versions of NORK-IC (NORK-ICÄ1 and NORK-ICÄ2) were generated with restriction enzymes. The Y2H cDNA library in pAD-GAL4 was made of young nodules and roots of M. truncatula line R108 (Györgyey et al., 2000). The truncated forms of Mt HMGR1ÄN, HMGR1ÄNÄC1, and HMGR1ÄNÄC2 were created by PCR coupled with oligos for restriction enzyme cloning. Screening was performed according the manufacturer's procedure (Stratagene). Colonies were collected and plated on SD-WLHA medium. Similar transformation and selection procedures were used for direct interaction studies.

Transfection of Arabidopsis Protoplasts and Protein Extraction

For transient expression of proteins, HMGR1ÄN, NORK-IC, and GFP were cloned under the control of the 35S promoter in the pRT104 vector (Topfer et al., 1987) and tagged with the HA or c-Myc epitopes. A cell suspension culture derived from Arabidopsis thaliana ecotype Columbia seedlings was grown in Murashige and Skoog (MS) complete liquid media containing sucrose (30 g/L), kinetin (14 μg/L), and 2,4-D at 23°C with 135 rpm shaking under continuous light and weekly subcultured (15 mL of culture in 85 mL of fresh medium). Forty milliliters of a 5-d-old cell suspension were treated with cell wall digestion solution (0.1 g/mL of cellulase Serva R10, 0.02 g/mL of macerozyme Yakult, 4.13 g/L of MS, 0.34 M glucose, and 0.34 M mannitol, pH 5.5) overnight at room temperature in the dark.

Cells were centrifuged and the pellet was washed with culturing medium (4.13 g/L of MS, 0.34 M glucose, and 0.34 M mannitol, pH 5.5). Protoplasts were then separated from the debris using a density gradient (4.13 g/L MS and 30 g/L sucrose, pH 5.5) coupled to centrifugation. For each transformation, 15 μg of plasmid DNA was added to 10 protoplasts, and transformation was performed by 20 min of incubation with polyethylene glycol solution (25% [w/v] PEG 6000, 0.45 M mannitol, and 0.1 M calcium nitrate, pH 9) in the dark. Finally, protoplasts were rinsed with 0.275 M $Ca(NO_3)_2$ and incubated overnight in the dark in the culturing medium.

Protoplasts were then pelleted and resuspended in extraction buffer (25 mM Tris-HCl, pH 7.7, 10 mM $MgCl_2$, 75 mM NaCl, 15 mM â-glycerophosphate, 0.1% Tween 20, 10% glycerol, 1 mM DTT, 1 mM NaF, 0.5 mM $NaVO_3$, and 0.5 mM PMSF), vortexed, and frozen in liquid nitrogen. Soluble proteins were separated from the debris by centrifugation at 4°C.

Coimmunoprecipitation and Protein Gel Blot Analysis

For each coimmunoprecipitation reaction, 300 µL of total protein extract at 5 µg/µL were mixed to 500 ng of rat anti-HA antibody (clone 3F10; Roche) and 80 µL of magnetic micro beads bound to G-protein (MACS Protein G Microbeads; Amersham). The mixture was agitated overnight at 4°C. Coimmunoprecipitations were performed on MACS columns (Amersham) as suggested by the manufacturer. The different protein fractions were separated on 10% SDS-PAGE in electrophoresis buffer (25 mM Tris-HCl, pH 8.3, 250 mM glycine, and 0.1% SDS) and then transferred to PVDF membranes (Amersham) in transfer buffer (48 mM Tris-HCl, pH 8.3, 39 mM glycine, and 20% ethanol). Membranes were then saturated with blocking solution (50 mM Tris-HCl, pH 7.4, 25 mM NaCl, 25 mM KCl, 0.05% Tween 20, and 5% defatted milk powder). The mouse anti-HA antibody (12CA5; Boehringer Mannheim) or mouse anti-c-Myc antibody (9E10; Boehringer-Mannheim) were applied at 0.08 µg/mL into the blocking solution overnight with slow shaking at 4°C. Secondary sheep antibody coupled to horseradish peroxidase (IgG anti-mouse A-5906; Sigma-Aldrich) was diluted 1:30,000 into the blocking solution and applied to the membrane. Chemoluminescence was finally detected with the commercial ECL detection kit (Amersham).

Subcellular Localization of Mt HMGR1 in Arabidopsis Protoplasts

The full-length Mt HMGR1 open reading frame, fused to the YFP open reading frame, was cloned in pRT104 under the control of 35S promoter. The fusion protein was transiently expressed in Arabidopsis protoplasts as described above. Three days after transfection, protoplasts were observed by confocal microscopy.

RT-PCR and in Situ Hybridization

In three independent experiments, total RNA was isolated from M. truncatula (Jemalong) roots, nodules, leaves, stems, and flowers by the RNeasy plant mini kit (Qiagen) for cDNA synthesis. To remove traces of genomic DNA, equal amounts of total RNA were treated with DNase (FPLC pure; Amersham) that was subsequently heat inactivated. cDNAs were synthesized by retrotranscription in the presence of Powerscript reverse transcriptase (Clontech), RNase inhibitor (RNasin; Promega), and oligo(dT) primers. Diluted fractions were used in PCR (EUROBIO TAQ polymerase). HMGR and NORK cDNAs were amplified in 35 cycles (94°C for 30 s, 55°C for 30 s, and 72°C for 1 min). Mtc27 constitutive marker was amplified in 25 cycles. PCR products were analyzed by the Fisher Scientific Bioblock gel documentation system.

In situ hybridizations were performed according to de Almeida Engler et al. (2001). DIG-labeled specific antisense or sense riboprobes were generated for Mt HMGR1 (226 bp from the 5' untranslated region) and for Mt NORK

(381-bp fragment). Hybridization signals were revealed with DIG antibodies and alkaline phosphatase–conjugated secondary antibodies. Signals were observed with a light microscope and CCD camera.

RNAi

The 5' untranslated region with a short part of the coding region was amplified from Mt HMGR1, resulting in a 226-bp PCR fragment (same fragment as for the in situ hybridization). As a control, the corresponding fragment was amplified from Mt HMGR4, generating a PCR product of 158 bp. The PCR products were first cloned in pENTR/D TOPO (Invitrogen). LR recombination was performed with the modified pK7GWIWG2(II) binary vector containing the constitutively expressed fluorescent DsRED1 marker (Limpens et al., 2004). Agrobacterium rhizogenes MSU440 was transformed with these constructs and the empty vector. A. rhizogenes–mediated hairy root transformation was done according to Boisson-Dernier et al. (2001). DsRED1 fluorescence in the roots was observed with confocal microscopy.

Mapping Genes in Microorganisms

As with any type of map, a genetic map must show the positions of distinctive features. In a geographic map these markers are recognizable components of the landscape, such as rivers, roads and buildings. What markers can we use in a genetic landscape?

Genes were the First Markers to be Used

The first genetic maps, constructed in the early decades of the 20th century for organisms such as the fruit fly, used genes as markers. This was many years before it was understood that genes are segments of DNA molecules. Instead, genes were looked upon as abstract entities responsible for the transmission of heritable characteristics from parent to offspring. To be useful in genetic analysis, a heritable characteristic has to exist in at least two alternative forms or phenotypes, an example being tall or short stems in the pea plants originally studied by Mendel. Each phenotype is specified by a different allele of the corresponding gene. To begin with, the only genes that could be studied were those specifying phenotypes that were distinguishable by visual examination. So, for example, the first fruit-fly maps showed the positions of genes for body color, eye color, wing shape and suchlike, all of these phenotypes being visible simply by looking at the flies with a low-power microscope or the naked eye. This approach was fine in the early days but geneticists soon realized that there were only a limited number of visual phenotypes whose inheritance could be studied, and in many cases their analysis was complicated because a single phenotype could be affected by more than one gene. For example, by 1922 over 50 genes had been mapped onto the four fruit-fly chromosomes, but nine of these were for eye color; in later research, geneticists studying fruit flies had to learn to distinguish between fly eyes that were colored red, light red, vermilion, garnet, carnation, cinnabar, ruby, sepia, scarlet, pink, cardinal, claret, purple or brown. To make gene maps more comprehensive it would be necessary to find characteristics that were more distinctive and less complex than visual ones.

The answer was to use biochemistry to distinguish phenotypes. This has been particularly important with two types of organisms - microbes and humans. Microbes, such as bacteria and yeast, have very few visual characteristics so gene mapping with these organisms has to rely on biochemical phenotypes such as those listed in . With humans it is possible to use visual characteristics, but since the 1920s studies of human genetic variation have been based largely on biochemical phenotypes that can be scored by blood typing. These phenotypes include not only the standard blood groups such as the ABO series (Yamamoto et al., 1990), but also variants of blood serum proteins and of immunological proteins such as the human leukocyte antigens (the HLA system). A big advantage of these markers is that many of the relevant genes have multiple alleles. For example, the gene called HLA-DRB1 has at least 290 alleles and HLA-B has over 400. This is relevant because of the way in which gene mapping is carried out with humans (Section 5.2.4).

Rather than setting up many breeding experiments, which is the procedure with experimental organisms such as fruit flies or mice, data on inheritance of human genes have to be gleaned by examining the phenotypes displayed by members of a single family. If all the family members have the same allele for the gene being studied then no useful information can be obtained. It is therefore necessary for the relevant marriages to have occurred, by chance, between individuals with different alleles. This is much more likely if the gene being studied has 290 rather than two alleles.

DNA Markers for Genetic Mapping

Genes are very useful markers but they are by no means ideal. One problem, especially with larger genomes such as those of vertebrates and flowering plants, is that a map based entirely on genes is not very detailed. This would be true even if every gene could be mapped. In most eukaryotic genomes the genes are widely spaced out with large gaps between them.

The problem is made worse by the fact that only a fraction of the total number of genes exist in allelic forms that can be distinguished conveniently. Gene maps are therefore not very comprehensive. We need other types of marker. Mapped features that are not genes are called DNA markers. As with gene markers, a DNA marker must have at least two alleles to be useful. There are three types of DNA sequence feature that satisfy this requirement: restriction fragment length polymorphisms (RFLPs), simple sequence length polymorphisms (SSLPs), and single nucleotide polymorphisms (SNPs).

Restriction Fragment Length Polymorphisms (RFLPs)

RFLPs were the first type of DNA marker to be studied. Recall that restriction enzymes cut DNA molecules at specific recognition sequences (Section 4.1.2). This sequence specificity means that treatment of a DNA molecule with a restriction enzyme should always produce the same set of fragments.

This is not always the case with genomic DNA molecules because some restriction sites are polymorphic, existing as two alleles, one allele displaying the correct sequence for the restriction site and therefore being cut when the DNA is treated with the enzyme, and the second allele having a sequence alteration so the restriction site is no longer recognized. The result of the sequence alteration is that the two adjacent restriction fragments remain linked together after treatment with the enzyme, leading to a length polymorphism.

This is an RFLP and its position on a genome map can be worked out by following the inheritance of its alleles, just as is done when genes are used as markers. There are thought to be about 10 RFLPs in the human genome, but of course for each RFLP there can only be two alleles (with and without the site). The value of RFLPs in human gene mapping is therefore limited by the high possibility that the RFLP being studied shows no variability among the members of an interesting family.

A restriction fragment length polymorphism (RFLP). The DNA molecule on the left has a polymorphic restriction site (marked with the asterisk) that is not present in the molecule on the right. The RFLP is revealed after treatment with the restriction enzyme.

In order to score an RFLP, it is necessary to determine the size of just one or two individual restriction fragments against a background of many irrelevant fragments. This is not a trivial problem: an enzyme such as EcoRI, with a 6-bp recognition sequence, should cut approximately once every 4 = 4096 bp and so would give almost 800 000 fragments when used with human DNA.

After separation by agarose gel electrophoresis, these 800 000 fragments produce a smear and the RFLP cannot be distinguished. Southern hybridization, using a probe that spans the polymorphic restriction site, provides one way of visualizing the RFLP, but nowadays PCR is more frequently used.

The primers for the PCR are designed so that they anneal either side of the polymorphic site, and the RFLP is typed by treating the amplified fragment with the restriction enzyme and then running a sample in an agarose gel.

Two methods for scoring an RFLP. (A) RFLPs can be scored by Southern hybridization. The DNA is digested with the appropriate restriction enzyme and separated in an agarose gel. The smear of restriction fragments is transferred to a nylon membrane.

Simple Sequence Length Polymorphisms (SSLPs)

SSLPs are arrays of repeat sequences that display length variations, different alleles containing different numbers of repeat units. Unlike RFLPs, SSLPs can be multi-allelic as each SSLP can have a number of different length variants. There are two types of SSLP, both of which were described in Section 2.4.1:

- Minisatellites, also known as variable number of tandem repeats (VNTRs), in which the repeat unit is up to 25 bp in length;
- Microsatellites or simple tandem repeats (STRs), whose repeats are shorter, usually dinucleotide or tetranucleotide units.

Microsatellites are more popular than minisatellites as DNA markers, for two reasons. First, minisatellites are not spread evenly around the genome but tend to be found more frequently in the telomeric regions at the ends of chromosomes. In geographic terms, this is equivalent to trying to use a map of lighthouses to find one's way around the middle of an island. Microsatellites are more conveniently spaced throughout the genome. Second, the quickest way to type a length polymorphism is by PCR, but PCR typing is much quicker and more accurate with sequences less than 300 bp in length. Most minisatellite alleles are longer than this because the repeat units are relatively large and there tend to be many of them in a single array, so PCR products of several kb are needed to type them. Typical microsatellites consist of 10–30 copies of a repeat that is usually no longer than 4 bp in length, and so are much more amenable to analysis by PCR. There are 6.5 × 10 microsatellites in the human genome.

Single nucleotide polymorphisms (SNPs)

These are positions in a genome where some individuals have one nucleotide (e.g. a G) and others have a different nucleotide (e.g. a C). There are vast numbers of SNPs in every genome, some of which also give rise to RFLPs, but many of which do not because the sequence in which they lie is not recognized by any restriction enzyme. In the human genome there are at least 1.42 million SNPs, only 100 000 of which result in an RFLP (SNP Group, 2001).

A Single Nucleotide Polymorphism (SNP)

Although each SNP could, potentially, have four alleles (because there are four nucleotides), most exist in just two forms, so these markers suffer from the same drawback as RFLPs with regard to human genetic mapping: there is a high possibility that a SNP does not display any variability in the family that is being studied. The advantages of SNPs are their abundant numbers and the fact that they can be typed by methods that do not involve gel electrophoresis. This is important because gel electrophoresis has proved difficult to automate so any detection method that uses it will be relatively slow and labor-intensive.

SNP detection is more rapid because it is based on oligonucleotide hybridization analysis. An oligonucleotide is a short single-stranded DNA molecule, usually less than 50 nucleotides in length, that is synthesized in the test tube. If the conditions are just right, then an oligonucleotide will hybridize with another DNA molecule only if the oligonucleotide forms a completely base-paired structure with the second molecule. If there is a single mismatch - a single position within the oligonucleotide that does not form a base pair - then hybridization does not occur. Oligonucleotide hybridization can therefore discriminate between the two alleles of an SNP. Various screening strategies have been devised (Mir and Southern, 2000), including DNA chip technology and solution hybridization techniques.

Oligonucleotide hybridization is very specific. Under highly stringent hybridization conditions, a stable hybrid occurs only if the oligonucleotide is able to form a completely base-paired structure with the target DNA. If there is a single mismatch. DNA microarrays and chips. High-density arrays of DNA molecules for parallel hybridization analyses. DNA microarrays and chips are designed to allow many hybridization experiments to be performed in parallel. Their main applications have been in the screening.

- A DNA chip is a wafer of glass or silicon, 2.0 cm or less in area, carrying many different oligonucleotides in a high-density array. The DNA to be tested is labeled with a fluorescent marker and pipetted onto the surface of the chip. Hybridization is detected by examining the chip with a fluorescence microscope, the positions at which the fluorescent signal is emitted indicating which oligonucleotides have hybridized with the test DNA. Many SNPs can therefore be scored in a single experiment (Wang et al., 1998; Gerhold et al., 1999).
- Solution hybridization techniques are carried out in the wells of a microtiter tray, each well containing a different oligonucleotide, and use a detection system that can discriminate between unhybridized single-stranded DNA and the double-stranded product that results when an oligonucleotide hybridizes to the test DNA. Several systems have been developed, one of which makes use of a pair of labels comprising a fluorescent dye and a compound that quenches the fluorescent signal when brought into close proximity with the dye. The dye is attached to one end of an oligonucleotide and the quenching compound to the other end. Normally there is no fluorescence because the oligonucleotide is designed in such a way that the two ends base-pair to one another, placing the quencher next to the dye. Hybridization between oligonucleotide and test DNA disrupts this base pairing, moving the quencher away from the dye and enabling the fluorescent signal to be generated (Tyagi et al., 1998).

One way of detecting an SNP by solution hybridization. The oligonucleotide probe has two end-labels. One of these is a fluorescent dye and the other is a quenching compound. The two ends of the oligonucleotide base-pair to one another, so the fluorescent.

Why do SNPs have only two alleles?. Any of the four nucleotides could be present at any position in the genome, so it might be imagined that each single nucleotide polymorphism (SNP) should have four alleles. Theoretically this is possible but in practice.

Linkage Analysis is the Basis of Genetic Mapping

Now that we have assembled a set of markers with which to construct a genetic map we can move on to look at the mapping techniques themselves. These techniques are all based on genetic linkage, which in turn derives from the seminal discoveries in genetics made in the mid 19th century by Gregor Mendel.

The Principles of Inheritance and the discovery of linkage

Genetic mapping is based on the principles of inheritance as first described by Gregor Mendel in 1865 (Orel, 1995). From the results of his breeding experiments with peas, Mendel concluded that each pea plant possesses two alleles for each gene, but displays only one phenotype. This is easy to understand if the plant is pure-breeding, or homozygous, for a particular characteristic, as it then possesses two identical alleles and displays the appropriate phenotype.

However, Mendel showed that if two pure-breeding plants with different phenotypes are crossed then all the progeny (the F_1 generation) display the same phenotype.

These F_1 plants must be heterozygous, meaning that they possess two different alleles, one for each phenotype, one allele inherited from the mother and one from the father. Mendel postulated that in this heterozygous condition one allele overrides the effects of the other allele; he therefore described the phenotype expressed in the F_1 plants as being dominant over the second, recessive phenotype. This is the perfectly correct interpretation of the interaction between the pairs of alleles studied by Mendel, but we now appreciate that this simple dominant-recessive rule can be complicated by situations that he did not encounter. One of these is incomplete dominance, where the heterozygous phenotype is intermediate between the two homozygous forms. An example is when red carnations are crossed with white ones, the F_1 heterozygotes being pink. Another complication is codominance, when both alleles are detectable in the heterozygote. Codominance is the typical situation for DNA markers.

Homozygosity and heterozygosity. Mendel studied seven pairs of contrasting characteristics in his pea plants, one of which was violet and white flower color, as shown here. (A) Pure-breeding plants always give rise to flowers with the parental color.

As well as discovering dominance and recessiveness, Mendel carried out additional crosses that enabled him to establish two Laws of Genetics. The First Law states that alleles segregate randomly. In other words, if the parent's alleles are A and a, then a member of the F_1 generation has the same chance of inheriting A as it has of inheriting. The Second Law is that pairs of alleles segregate independently, so that inheritance of the alleles of gene A is independent of inheritance of the alleles of gene B. Because of these laws, the outcomes of genetic crosses are predictable. Mendel's Laws enable the outcome of genetic crosses to be predicted. (A) Mendel's First Law states that alleles segregate randomly. The example shows inheritance of alleles A and a in a cross involving two heterozygous parents. Each member of the F_1 generation.

When Mendel's work was rediscovered in 1900, his Second Law worried the early geneticists because it was soon established that genes reside on chromosomes, and it was realized that all organisms have many more genes than chromosomes. Chromosomes are inherited as intact units, so it was reasoned that the alleles of some pairs of genes will be inherited together because they are on the same chromosome. This is the principle of genetic linkage, and it was quickly shown to be correct, although the results did not turn out exactly as expected. The complete linkage that had been anticipated between many pairs of genes failed to materialize. Pairs of genes were either inherited independently, as expected for genes in different chromosomes, or, if they showed linkage, then it was only partial linkage: sometimes they were inherited together and sometimes they were not. The resolution of this contradiction between theory and observation was the critical step in the development of genetic mapping techniques.

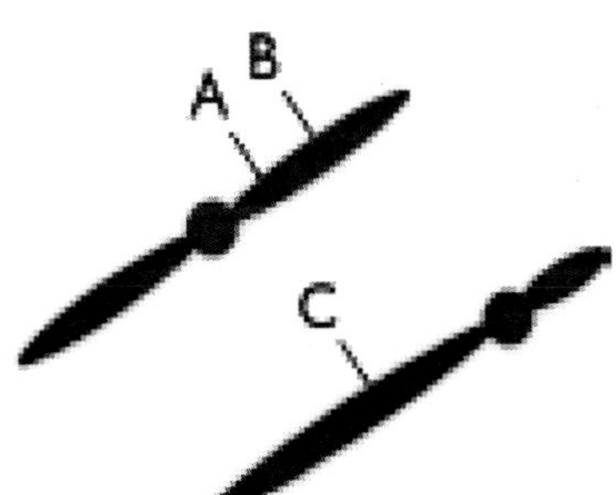

Genes on the same chromosome should display linkage. Genes A and B are on the same chromosome and so should be inherited together. Mendel's Second Law should therefore not apply to the inheritance of A and B, but holds for the inheritance of A and C.

Partial linkage. Partial linkage was discovered in the early 20th century. The cross shown here was carried out by Bateson, Saunders and Punnett in 1905 with sweet peas. The parental cross gives the typical dihybrid result. Partial linkage is explained by the behavior of chromosomes during meiosis.

The critical breakthrough was achieved by Thomas Hunt Morgan, who made the conceptual leap between partial linkage and the behavior of chromosomes when the nucleus of a cell divides. Cytologists in the late 19th century had distinguished two types of nuclear division: mitosis and meiosis. Mitosis is more common, being the process by which the diploid nucleus of a somatic cell divides to produce two daughter nuclei, both of which are diploid. Approximately 10 mitoses are needed to produce all the cells required during a human lifetime. Before mitosis begins, each chromosome in the nucleus is replicated, but the resulting daughter chromosomes do not immediately break away from one another. To begin with they remain attached at their centromeres and by cohesin proteins which act as 'molecular glue' holding together the arms of the replicated chromosomes. The daughters do not separate until later in mitosis when the chromosomes are distributed between the two new nuclei. Obviously it is important that each of the new nuclei receives a complete set of chromosomes, and most of the intricacies of mitosis appear to be devoted to achieving this end.

Mitosis. During interphase (the period between nuclear divisions) the chromosomes are in their extended form (Section 2.2.1). At the start of mitosis the chromosomes condense and by late prophase have formed structures that are visible with the light.

Mitosis illustrates the basic events occurring during nuclear division but is not directly relevant to genetic mapping. Instead, it is the distinctive features of meiosis that interest us. Meiosis occurs only in reproductive cells, and results in a diploid cell giving rise to four haploid gametes, each of which can subsequently fuse with a gamete of the opposite sex during sexual reproduction. The fact that meiosis results in four haploid cells whereas mitosis gives rise to two diploid cells is easy to explain: meiosis involves two nuclear divisions, one after the other, whereas mitosis is just a single nuclear division. This is an important distinction, but the critical difference between mitosis and meiosis is more subtle.

Recall that in a diploid cell there are two separate copies of each chromosome. We refer to these as pairs of homologous chromosomes. During mitosis, homologous chromosomes remain separate from one another, each member of the pair replicating and being passed to a daughter nucleus independently of its homolog. In meiosis, however, the pairs of homologous chromosomes are by no means independent. During meiosis I, each chromosome lines up with its homolog to form a bivalent. This occurs after each chromosome has replicated, but before the replicated structures split, so the bivalent in fact contains four chromosome copies, each of which is destined to find its way into one of the four gametes that will be produced at the end of the meiosis. Within the bivalent, the chromosome arms (the chromatids) can undergo physical breakage and exchange of segments of DNA. The process is called crossing-over or recombination and was discovered by the Belgian cytologist Janssens in 1909. This was just 2 years before Morgan started to think about partial linkage. Meiosis. The events involving one pair of homologous chromosomes are shown; one member of the pair is red, the other is blue. At the start of meiosis the chromosomes condense and each homologous pair lines up to form a bivalent. Within the bivalent, crossing-over.

How did the discovery of crossing-over help Morgan explain partial linkage? To understand this we need to think about the effect that crossing-over can have on the inheritance of genes. Let us consider two genes, each of which has two alleles. We will call the first gene A and its alleles A and a, and the second gene B with alleles B and b.

Imagine that the two genes are located on chromosome number 2 of Drosophila melanogaster, the species of fruit fly studied by Morgan. We are going to follow the meiosis of a diploid nucleus in which one copy of chromosome 2 has alleles A and B, and the second has a and b.

The effect of a crossover on linked genes. The drawing shows a pair of homologous chromosomes, one red and the other blue. A and B are linked genes with alleles A, a, B and b. On the left is a meiosis with no crossover between A and B: two of the resulting.

1. A crossover does not occur between genes A and B. If this is what happens then two of the resulting gametes will contain chromosome copies with alleles A and B, and the other two will contain a and b. In other words, two of the gametes have the genotype AB and two have the genotype ab.
2. A crossover does occur between genes A and B. This leads to segments of DNA containing gene B being exchanged between homologous chromosomes. The eventual result is that each gamete has a different genotype: 1 AB, 1 aB, 1 Ab, 1 ab.

Now think about what would happen if we looked at the results of meiosis in a hundred identical cells. If crossovers never occur then the resulting gametes will have the following genotypes:

$$200AB$$

$$200ab$$

This is complete linkage: genes A and B behave as a single unit during meiosis. But if (as is more likely) crossovers occur between A and B in some of the nuclei, then the allele pairs will not be inherited as single units. Let us say that crossovers occur during 40 of the 100 meioses. The following gametes will result:

160 *AB*

160 *ab*

40 *Ab*

40 *aB*

The linkage is not complete, it is only partial. As well as the two parental genotypes (AB, ab) we see gametes with recombinant genotypes (Ab, aB).

From Partial Linkage to Genetic Mapping

Once Morgan had understood how partial linkage could be explained by crossing-over during meiosis he was able to devise a way of mapping the relative positions of genes on a chromosome. In fact the most important work was done not by Morgan himself, but by an undergraduate in his laboratory, Arthur Sturtevant (Sturtevant, 1913). Sturtevant assumed that crossing-over was a random event, there being an equal chance of it occurring at any position along a pair of lined-up chromatids. If this assumption is correct then two genes that are close together will be separated by crossovers less frequently than two genes that are more distant from one another. The frequency with which the genes are unlinked by crossovers will be directly proportional to how far apart they are on their chromosome. The recombination frequency is therefore a measure of the distance between two genes. If you work out the recombination frequencies for different pairs of genes, you can construct a map of their relative positions on the chromosome. Working out a genetic map from recombination frequencies. The example is taken from the original experiments carried out with fruit flies by Arthur Sturtevant. All four genes are on the X chromosome of the fruit fly. Recombination frequencies between.

It turns out that Sturtevant's assumption about the randomness of crossovers was not entirely justified. Comparisons between genetic maps and the actual positions of genes on DNA molecules, as revealed by physical mapping and DNA sequencing, have shown that some regions of chromosomes, called recombination hotspots, are more likely to be involved in crossovers than others. This means that a genetic map distance does not necessarily indicate the physical distance between two markers. Also, we now realize that a single chromatid can participate in more than one crossover at the same time, but that there are limitations on how close together these crossovers can be, leading to more inaccuracies in the mapping procedure. Despite these qualifications, linkage analysis usually makes correct deductions about gene order, and distance estimates are sufficiently accurate to generate genetic maps that are of value as frameworks for genome sequencing projects. Comparison between the genetic and physical maps of Saccharomyces cerevisiae chromosome III. The comparison shows the discrepancies between the genetic and physical maps, the latter determined by DNA sequencing. Note that the order of the upper two markers.

Linkage Analysis with Different Types of Organism

To see how linkage analysis is actually carried out, we need to consider three quite different situations:

- Linkage analysis with species such as fruit flies and mice, with which we can carry out planned breeding experiments;
- Linkage analysis with humans, with whom we cannot carry out planned experiments but instead make use of family pedigrees;
- Linkage analysis with bacteria, which do not undergo meiosis.

Linkage analysis when planned breeding experiments are possible

The first type of linkage analysis is the modern counterpart of the method developed by Morgan and his colleagues. The method is based on analysis of the progeny of experimental crosses set up between parents of known genotypes and is, at least in theory, applicable to all eukaryotes. Ethical considerations preclude this approach in humans, and practical problems such as the length of the gestation period and the time taken for the newborn to reach maturity (and hence to participate in subsequent crosses) limit the effectiveness of the method with some animals and plants. We see that the key to gene mapping is being able to determine the genotypes of the gametes resulting from meiosis. In a few situations this is possible by directly examining the gametes. For example, the gametes produced by some microbial eukaryotes, including the yeast Saccharomyces cerevisiae, can be grown into colonies of haploid cells, whose genotypes can be determined by biochemical tests. Direct genotyping of gametes is also possible with higher eukaryotes if DNA markers are used, as PCR can be carried out with the DNA from individual spermatozoa, enabling RFLPs, SSLPs and SNPs to be typed. Unfortunately, sperm typing is laborious.

Routine linkage analysis with higher eukaryotes is therefore carried out not by examining the gametes directly but by determining the genotypes of the diploid progeny that result from fusion of two gametes, one from each of a pair of parents. In other words, a genetic cross is performed.

The complication with a genetic cross is that the resulting diploid progeny are the product not of one meiosis but of two (one in each parent), and in most organisms crossover events are equally likely to occur during production of the male and female gametes. Somehow we have to be able to disentangle from the genotypes of the diploid progeny the crossover events that occurred in each of these two meioses. This means that the cross has to be set up with care. The standard procedure is to use a test cross.

Scenario 1, where we have set up a test cross to map the two genes we met earlier: gene A (alleles A and a) and gene B (alleles B and b), both on chromosome 2 of the fruit fly. The critical feature of a test cross is the genotypes of the two parents:

- One parent is a double heterozygote. This means that all four alleles are present in this parent: its genotype is AB/ab. This notation indicates that one pair of the homologous chromosomes has alleles A and B, and the other has a and b. Double heterozygotes can be obtained by crossing two pure-breeding strains, for example AB/AB × ab/ab.
- The second parent is a pure-breeding double homozygote. In this parent both homologous copies of chromosome 2 are the same: in the example shown in Scenario 1 both have alleles a and b and the genotype of the parent is ab/ab.

The double heterozygote has the same genotype as the cell whose meiosis. Our objective is therefore to infer the genotypes of the gametes produced by this parent and to calculate the fraction that are recombinants. Note that all the gametes produced by the second parent (the double homozygote) will have the genotype ab regardless of whether they are parental or recombinant gametes. Alleles a and b are both recessive, so meiosis in this parent is, in effect, invisible when the genotypes of the progeny are examined. This means that, as shown in Scenario 1, the genotypes of the diploid progeny can be unambiguously converted into the genotypes of the gametes from the double heterozygous parent. The test cross therefore enables us to make a direct examination of a single meiosis and hence to calculate a recombination frequency and map distance for the two genes being studied.

Just one additional point needs to be considered. If, as in Scenario 1, gene markers displaying dominance and recessiveness are used, then the double homozygous parent must have alleles for the two recessive phenotypes; however, if codominant DNA markers are used, then the double homozygous parent can have any combination of homozygous alleles (i.e. AB/AB, Ab/Ab, aB/aB and ab/ab). Scenario 2. Multipoint crosses. The power of linkage analysis is enhanced if more than two markers are followed in a single cross. Not only does this generate recombination frequencies more quickly, but it also enables the relative order of markers on a chromosome.

Gene Mapping by Human Pedigree Analysis

With humans it is of course impossible to pre-select the genotypes of parents and set up crosses designed specifically for mapping purposes. Instead, data for the calculation of recombination frequencies have to be obtained by examining the genotypes of the members of successive generations of existing families.

This means that only limited data are available, and their interpretation is often difficult because a human marriage rarely results in a convenient test cross, and often the genotypes of one or more family members are unobtainable because those individuals are dead or unwilling to cooperate.

In this example we are studying a genetic disease present in a family of two parents and six children. Genetic diseases are frequently used as gene markers in humans, the disease state being one allele and the healthy state being a second allele. The mother is affected by the disease, as are four of her children. We know from family accounts that the maternal grandmother also suffered from this disease, but both she and her husband - the maternal grandfather - are now dead. We can include them in the pedigree, with slashes indicating that they are dead, but we cannot obtain. Our aim is to map the position of the gene for the genetic disease. For this purpose we are studying its linkage to a microsatellite marker M, four alleles of which - M_1, M_2, M_3 and M_4 - are present in the living family members. The question is, how many of the children are recombinants?

If we look at the genotypes of the six children we see that numbers 1, 3 and 4 have the disease allele and the microsatellite allele M_1. Numbers 2 and 5 have the healthy allele and M_2. We can therefore construct two alternative hypotheses. The first is that the two copies of the relevant homologous chromosomes in the mother have the genotypes Disease-M_1 and Healthy-M_2; therefore children 1, 2, 3, 4 and 5 have parental genotypes and child 6 is the one and only recombinant. This would suggest that the disease gene and the microsatellite are relatively closely linked and that crossovers between them occur infrequently. The alternative hypothesis is that the mother's chromosomes have the genotypes Healthy-M_1 and Disease-M_2; this would mean that children 1–5 are recombinants, and child 6 has the parental genotype. This would mean that the gene and microsatellite are relatively far apart on the chromosome. We cannot determine which of these hypotheses is correct: the data are frustratingly ambiguous.

The most satisfying solution to the problem posed by the pedigree would be to know the genotype of the grandmother. Let us pretend that this is a soap opera family and that the grandmother is not really dead. To everyone's surprise she reappears just in time to save the declining audience ratings. Her genotype for microsatellite M turns out to be M_1M_5. This tells us that the disease allele is on the same chromosome as M_1. We can therefore conclude with certainty that Hypothesis 1 is correct and that only child 6 is a recombinant.

Resurrection of key individuals is not usually an option open to real-life geneticists, although DNA can be obtained from old pathology specimens such as slides and Guthrie cards. Imperfect pedigrees are analyzed statistically, using a measure called the lod score (Morton, 1955). This stands for logarithm of the odds that the genes are linked and is used primarily to determine if the two markers being studied lie on the same chromosome, in other words if the genes are linked or not. If the lod analysis establishes linkage then it can also provide a measure of the most likely recombination frequency. Ideally the available data will derive from more than one pedigree, increasing the confidence in the result. The analysis is less ambiguous for families with larger numbers of children, and, as it is important that the members of at least three generations can be genotyped. For this reason, family collections have been established, such as the one maintained by the Centre d'Études du Polymorphisme Humaine (CEPH) in Paris (Dausset et al., 1990). The CEPH collection contains cultured cell lines from families in which all four grandparents as well as at least eight second-generation children could be sampled. This collection is available for DNA marker mapping by any researcher who agrees to submit the resulting data to the central CEPH database.

Genetic Mapping in Bacteria

The final type of genetic mapping that we must consider is the strategy used with bacteria. The main difficulty that geneticists faced when trying to develop genetic mapping techniques for bacteria is that these organisms are normally haploid, and so do not undergo meiosis. Some other way therefore had to be devised to induce crossovers between homologous segments of bacterial DNA. The answer was to make use of three natural methods that exist for transferring pieces of DNA from one bacterium to another.

Three ways of achieving DNA transfer between bacteria. (A) Conjugation can result in transfer of chromosomal or plasmid DNA from the donor bacterium to the recipient. Conjugation involves physical contact between the two bacteria, with transfer thought.

- In conjugation two bacteria come into physical contact and one bacterium (the donor) transfers DNA to the second bacterium (the recipient).

  The transferred DNA can be a copy of some or possibly all of the donor cell's chromosome, or it could be a segment of chromosomal DNA - up to 1 Mb in length - integrated in a plasmid (Section 2.1.2). The latter is called episome transfer.
- Transduction involves transfer of a small segment of DNA - up to 50 kb or so - from donor to recipient via a bacteriophage.
- In transformation the recipient cell takes up from its environment a fragment of DNA, rarely longer than 50 kb, released from a donor cell.

After transfer, a double crossover must occur so that the DNA from the donor bacterium is integrated into the recipient cell's chromosome. If this does not occur then the transferred DNA is lost when the recipient cell divides. The only exception is after episome transfer, plasmids being able to propagate independently of the host chromosome. The basis of gene mapping in bacteria. (A) Transfer of a functional gene for tryptophan biosynthesis from a wild-type bacterium (genotype described as trp) to a recipient that lacks a functional copy of this gene (trp). The recipient.

Biochemical markers are invariably used, the dominant or wild-type phenotype being possession of a biochemical characteristic (e.g. ability to synthesize tryptophan) and the recessive phenotype being the complementary characteristic (e.g. inability to synthesize tryptophan).

The gene transfer is usually set up between a donor strain that possesses the wild-type alleles and a recipient with the recessive alleles, transfer into the recipient strain being monitored by looking for acquisition of the biochemical function(s) specified by the genes being studied.

The precise details of the mapping procedure depend on the type of gene transfer that is being used. In conjugation mapping the donor DNA is transferred as a continuous thread into the recipient, and gene positions are mapped by timing the entry of the wild-type alleles into the recipient.

Transduction and transformation mapping enable genes that are relatively close together to be mapped, because the transferred DNA segment is short (< 50 kb), so the probability of two genes being transferred together depends on how close together they are on the bacterial chromosome.

Physical Mapping

A map generated by genetic techniques is rarely sufficient for directing the sequencing phase of a genome project. This is for two reasons:

- The resolution of a genetic map depends on the number of crossovers that have been scored. This is not a major problem for microorganisms because these can be obtained in huge numbers, enabling many crossovers to be studied, resulting in a highly detailed genetic map in which the markers are just a few kb apart.

  For example, when the Escherichia coli genome sequencing project began in 1990, the latest genetic map for this organism comprised over 1400 markers, an average of one per 3.3 kb. This was sufficiently detailed to direct the sequencing program without the need for extensive physical mapping. Similarly, the Saccharomyces cerevisiae project was supported by a fine-scale genetic map (approximately 1150 genetic markers, on average one per 10 kb).

  The problem with humans and most other eukaryotes is that it is simply not possible to obtain large numbers of progeny, so relatively few meioses can be studied and the resolving power of linkage analysis is restricted. This means that genes that are several tens of kb apart may appear at the same position on the genetic map.

- Genetic maps have limited accuracy . We touched on this point in Section 5.2.3 when we assessed Sturtevant's assumption that crossovers occur at random along chromosomes. This assumption is only partly correct because the presence of recombination hotspots means that crossovers are more likely to occur at some points rather than at others. The effect that this can have on the accuracy of a genetic map was illustrated in 1992 when the complete sequence for S. cerevisiae chromosome III was published (Oliver et al., 1992), enabling the first direct comparison to be made between a genetic map and the actual positions of markers as shown by DNA sequencing.

There were considerable discrepancies, even to the extent that one pair of genes had been ordered incorrectly by genetic analysis. Bear in mind that S. cerevisiae is one of the two eukaryotes (fruit fly is the second) whose genomes have been subjected to intensive genetic mapping. If the yeast genetic map is inaccurate then how precise are the genetic maps of organisms subjected to less detailed analysis?

These two limitations of genetic mapping mean that for most eukaryotes a genetic map must be checked and supplemented by alternative mapping procedures before large-scale DNA sequencing begins. A plethora of physical mapping techniques has been developed to address this problem, the most important being:

- Restriction mapping, which locates the relative positions on a DNA molecule of the recognition sequences for restriction endonucleases;
- Fluorescent in situ hybridization (FISH), in which marker locations are mapped by hybridizing a probe containing the marker to intact chromosomes;
- Sequence tagged site (STS) mapping, in which the positions of short sequences are mapped by PCR and/or hybridization analysis of genome fragments.

Restriction Mapping

Genetic mapping using RFLPs as DNA markers can locate the positions of polymorphic restriction sites within a genome (Section 5.2.2), but very few of the restriction sites in a genome are polymorphic, so many sites are not mapped by this technique. Could we increase the marker density on a genome map by using an alternative method to locate the positions of some of the non-polymorphic restriction sites?

This is what restriction mapping achieves, although in practice the technique has limitations which mean that it is applicable only to relatively small DNA molecules. We will look first at the technique and then consider its relevance to genome mapping.

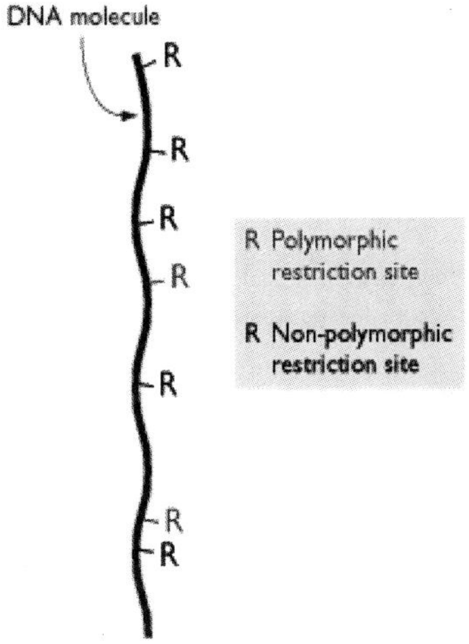

Figure: *Not All Restriction Sites are Polymorphic.*

The Basic Methodology for Restriction Mapping

The simplest way to construct a restriction map is to compare the fragment sizes produced when a DNA molecule is digested with two different restriction enzymes that recognize different target sequences. An example using the restriction enzymes EcoRI and BamHI. First, the DNA molecule is digested with just one of the enzymes and the sizes of the resulting fragments are measured by agarose gel electrophoresis.

Next, the molecule is digested with the second enzyme and the resulting fragments again sized in an agarose gel. The results so far enable the number of restriction sites for each enzyme to be worked out, but do not allow their relative positions to be determined. Additional information is therefore obtained by cutting the DNA molecule with both enzymes together.

Double restriction enables three of the sites to be mapped. However, a problem arises with the larger EcoRI fragment because this contains two BamHI sites and there are two alternative possibilities for the map location of the outer one of these. The problem is solved by going back to the original DNA molecule and treating it again with BamHI on its own, but this time preventing the digestion from going to completion by, for example, incubating the reaction for only a short time or using a suboptimal incubation temperature. This is called a partial restriction and leads to a more complex set of products, the complete restriction products now being supplemented with partially restricted fragments that still contain one or more uncut BamHI sites.

A partial restriction usually gives the information needed to complete a map, but if there are many restriction sites then this type of analysis becomes unwieldy, simply because there are so many different fragments to consider.

An alternative strategy is simpler because it enables the majority of the fragments to be ignored. This is achieved by attaching a radioactive or other type of marker to each end of the starting DNA molecule before carrying out the partial digestion. The result is that many of the partial restriction products become 'invisible' because they do not contain an end-fragment and so do not show up when the agarose gel is screened for labeled products. The sizes of the partial restriction products that are visible enable unmapped sites to be positioned relative to the ends of the starting molecule.

The Scale of Restriction Mapping is limited by the sizes of the restriction fragments. Restriction maps are easy to generate if there are relatively few cut sites for the enzymes being used. However, as the number of cut sites increases, so also do the numbers of single, double and partial restriction products whose sizes must be determined and compared in order for the map to be constructed. Computer analysis can be brought into play but problems still eventually arise. A stage will be reached when a digest contains so many fragments that individual bands merge on the agarose gel, increasing the chances of one or more fragments being measured incorrectly or missed out entirely. If several fragments have similar sizes then even if they can all be identified, it may not be possible to assemble them into an unambiguous map.

Restriction mapping is therefore more applicable to small rather than large molecules, with the upper limit for the technique depending on the frequency of the restriction sites in the molecule being mapped. In practice, if a DNA molecule is less than 50 kb in length it is usually possible to construct an unambiguous restriction map for a selection of enzymes with six-nucleotide recognition sequences. Fifty kb is of course way below the minimum size for bacterial or eukaryotic chromosomes, although it does cover a few viral and organelle genomes, and whole-genome restriction maps have indeed been important in directing sequencing projects with these small molecules. Restriction maps are equally useful after bacterial or eukaryotic genomic DNA has been cloned, if the cloned fragments are less than 50 kb, because a detailed restriction map can then be built up as a preliminary to sequencing the cloned region. This is an important application of restriction mapping in sequencing projects with large genomes, but is there any possibility of using restriction analysis for the more general mapping of entire genomes larger than 50 kb? The answer is a qualified 'yes', because the limitations of restriction mapping can be eased slightly by choosing enzymes expected to have infrequent cut sites in the target DNA molecule. These 'rare cutters' fall into two categories:

- Enzymes with seven- or eight-nucleotide recognition sequences . A few restriction enzymes cut at seven- or eight-nucleotide recognition sequences. Examples are SapI (52 -GCTCTTC-32) and SgfI (52 -GCGATCGC-32). The seven-nucleotide enzymes would be expected, on average, to cut a DNA molecule with a GC content of 50% once every 4 = 16 384 bp. The eight-nucleotide enzymes should cut once every 4 = 65 536 bp. Compare with 4 = 4096 bp for six-nucleotide enzymes such as BamHI and EcoRI. Seven- and eight-nucleotide cutters are often used in restriction mapping of large molecules but the approach is not as useful as it might be simply because not many of these enzymes are known.

- Enzymes whose recognition sequences contain motifs that are rare in the target DNA . Genomic DNA molecules do not have random sequences and some are significantly deficient in certain motifs. For example, the sequence 52 -CG-32 is rare in human DNA because human cells possess an enzyme that adds a methyl group to carbon 5 of the C nucleotide in this sequence. The resulting 5-methylcytosine is unstable and tends to undergo deamination to give thymine. The consequence is that during human evolution many of the 52 -CG-32 sequences that were originally in our genome have become converted to 52 -TG-32 . Restriction enzymes that recognize a site containing 52 -CG-32 therefore cut human DNA relatively infrequently. Examples are SmaI (52 -CCCGGG-32), which cuts human DNA on average once every 78 kb, and BssHII (52 -GCGCGC-32) which cuts once every 390 kb. Note that NotI, an eight-nucleotide cutter, also targets 52 -CG-32 sequences (recognition sequence 52 -GCGGCCGC-32) and cuts human DNA very rarely - approximately once every 10 Mb.

The potential of restriction mapping is therefore increased by using rare cutters. It is still not possible to construct restriction maps of the genomes of animals and plants, but it is feasible to use the technique with large cloned fragments, and the smaller DNA molecules of prokaryotes and lower eukaryotes such as yeast and fungi.

If a rare cutter is used then it may be necessary to employ a special type of agarose gel electrophoresis to study the resulting restriction fragments. This is because the relationship between the length of a DNA molecule and its migration rate in an electrophoresis gel is not linear, the resolution decreasing as the molecules get longer. This means that it is not possible to separate molecules more than about 50 kb in length because all of these longer molecules run as a single slowly migrating band in a standard agarose gel. To separate them it is necessary to replace the linear electric field used in conventional gel electrophoresis with a more complex field. An example is provided by orthogonal field alternation gel electrophoresis (OFAGE), in which the electric field alternates between two pairs of electrodes, each positioned at an angle of 45° to the length of the gel. The DNA molecules still move down through the gel, but each change in the field forces the molecules to realign. Shorter molecules realign more quickly than longer ones and so migrate more rapidly through the gel. The overall result is that molecules much longer than those separated by conventional gel electrophoresis can be resolved. Related techniques include CHEF (contour clamped homogeneous electric fields) and FIGE (field inversion gel electrophoresis).

Direct Examination of DNA Molecules for Restriction Sites

It is also possible to use methods other than electrophoresis to map restriction sites in DNA molecules. With the technique called optical mapping (Schwartz et al., 1993), restriction sites are directly located by looking at the cut DNA molecules with a microscope. The DNA must first be attached to a glass slide in such a way that the individual molecules become stretched out, rather than clumped together in a mass. There are two ways of doing this: gel stretching and molecular combing. To prepare gel-stretched DNA fibers (Schwartz et al., 1993), chromosomal DNA is suspended in molten agarose and placed on a microscope slide.

As the gel cools and solidifies, the DNA molecules become extended. To utilize gel stretching in optical mapping, the microscope slide onto which the molten agarose is placed is first coated with a restriction enzyme. The enzyme is inactive at this stage because there are no magnesium ions, which the enzyme needs in order to function. Once the gel has solidified it is washed with a solution containing magnesium chloride, which activates the restriction enzyme. A fluorescent dye is added, such as DAPI (4,6-diamino-2-phenylindole dihydrochloride), which stains the DNA so that the fibers can be seen when the slide is examined with a high-power fluorescence microscope. The restriction sites in the extended molecules gradually become gaps as the degree of fiber extension is reduced by the natural springiness of the DNA, enabling the relative positions of the cuts to be recorded.

In molecular combing (Michalet et al., 1997), the DNA fibers are prepared by dipping a silicone-coated cover slip into a solution of DNA, leaving it for 5 minutes (during which time the DNA molecules attach to the cover slip by their ends), and then removing the slip at a constant speed of 0.3 mm s. The force required to pull the DNA molecules through the meniscus causes them to line up. Once in the air, the surface of the cover slip dries, retaining the DNA molecules as an array of parallel fibers.

Optical mapping was first applied to large DNA fragments cloned in YAC and BAC vectors (Section 4.2.1). More recently, the feasibility of using this technique with genomic DNA has been proven with studies of a 1-Mb chromosome of the malaria parasite Plasmodium falciparum (Jing et al., 1999), and the two chromosomes and single megaplasmid of the bacterium Deinococcus radiodurans.

Fluorescent in Situ Hybridization (FISH)

The optical mapping method described above provides a link to the second type of physical mapping procedure that we will consider - FISH (Heiskanen et al., 1996). As in optical mapping, FISH enables the position of a marker on a chromosome or extended DNA molecule to be directly visualized. In optical mapping the marker is a restriction site and it is visualized as a gap in an extended DNA fiber. In FISH, the marker is a DNA sequence that is visualized by hybridization with a fluorescent probe.

In Situ Hybridization with Radioactive or Fluorescent Probes

In situ hybridization is a version of hybridization analysis (Section 4.1.2) in which an intact chromosome is examined by probing it with a labeled DNA molecule. The position on the chromosome at which hybridization occurs provides information about the map location of the DNA sequence used as the probe. For the method to work, the DNA in the chromosome must be made single stranded ('denatured') by breaking the base pairs that hold the double helix together. Only then will the chromosomal DNA be able to hybridize with the probe. The standard method for denaturing chromosomal DNA without destroying the morphology of the chromosome is to dry the preparation onto a glass microscope slide and then treat with formamide.

Fluorescent in situ hybridization. A sample of dividing cells is dried onto a microscope slide and treated with formamide so that the chromosomes become denatured but do not lose their characteristic metaphase morphologies.

In the early versions of in situ hybridization the probe was radioactively labeled but this procedure was unsatisfactory because it is difficult to achieve both sensitivity and resolution with a radioactive label, two critical requirements for successful in situ hybridization. Sensitivity requires that the radioactive label has a high emission energy (an example of such a radiolabel is P), but if the radiolabel has a high emission energy then it scatters its signal and so gives poor resolution. High resolution is possible if a radiolabel with low emission energy, such as H, is used, but these have such low sensitivity that lengthy exposures are needed, leading to a high background and difficulties in discerning the genuine signal.

These problems were solved in the late 1980s by the development of non-radioactive fluorescent DNA labels. These labels combine high sensitivity with high resolution and are ideal for in situ hybridization. Fluorolabels with different colored emissions have been designed, making it possible to hybridize a number of different probes to a single chromosome and distinguish their individual hybridization signals, thus enabling the relative positions of the probe sequences to be mapped. To maximize sensitivity, the probes must be labeled as heavily as possible, which in the past has meant that they must be quite lengthy DNA molecules - usually cloned DNA fragments of at least 40 kb. This requirement is less important now that techniques for achieving heavy labeling with shorter molecules have been developed. As far as the construction of a physical map is concerned, a cloned DNA fragment can be looked upon as simply another type of marker, although in practice the use of clones as markers adds a second dimension because the cloned DNA is the material from which the DNA sequence is determined. Mapping the positions of clones therefore provides a direct link between a genome map and its DNA sequence.

If the probe is a long fragment of DNA then one potential problem, at least with higher eukaryotes, is that it is likely to contain examples of repetitive DNA sequences (Section 2.4) and so may hybridize to many chromosomal positions, not just the specific point to which it is perfectly matched. To reduce this non-specific hybridization, the probe, before use, is mixed with unlabeled DNA from the organism being studied. This DNA can simply be total nuclear DNA (i.e. representing the entire genome) but it is better if a fraction enriched for repeat sequences is used. The idea is that the unlabeled DNA hybridizes to the repetitive DNA sequences in the probe, blocking these so that the subsequent in situ hybridization is driven wholly by the unique sequences (Lichter et al., 1990). Non-specific hybridization is therefore reduced or eliminated entirely.

FISH in Action

FISH was originally used with metaphase chromosomes (Section 2.2.1). These chromosomes, prepared from nuclei that are undergoing division, are highly condensed and each chromosome in a set takes up a recognizable appearance, characterized by the position of its centromere and the banding pattern that emerges after the chromosome preparation is stained. With metaphase chromosomes, a fluorescent signal obtained by FISH is mapped by measuring its position relative to the end of the short arm of the chromosome (the FLpter value). A disadvantage is that the highly condensed nature of metaphase chromosomes means that only low-resolution mapping is possible, two markers having to be at least 1 Mb apart to be resolved as separate hybridization signals (Trask et al., 1991). This degree of resolution is insufficient for the construction of useful chromosome maps, and the main application of metaphase FISH has been in determining the chromosome on which a new marker is located, and providing a rough idea of its map position, as a preliminary to finer scale mapping by other methods.

For several years these 'other methods' did not involve any form of FISH, but since 1995 a range of higher resolution FISH techniques has been developed. With these techniques, higher resolution is achieved by changing the nature of the chromosomal preparation being studied. If metaphase chromosomes are too condensed for fine-scale mapping then we must use chromosomes that are more extended. There are two ways of doing this (Heiskanen et al., 1996):

- Mechanically stretched chromosomes can be obtained by modifying the preparative method used to isolate chromosomes from metaphase nuclei. The inclusion of a centrifugation step generates shear forces which can result in the chromosomes becoming stretched to up to 20 times their normal length. Individual chromosomes are still recognizable and FISH signals can be mapped in the same way as with normal metaphase chromosomes. The resolution is significantly improved and markers that are 200–300 kb apart can be distinguished.
- Non-metaphase chromosomes can be used because it is only during metaphase that chromosomes are highly condensed: at other stages of the cell cycle the chromosomes are naturally unpacked. Attempts have been made to use prophase nuclei because in these the chromosomes are still sufficiently condensed for individual ones to be identified. In practice, however, these preparations provide no advantage over mechanically stretched chromosomes.

  Interphase chromosomes are more useful because this stage of the cell cycle (between nuclear divisions) is when the chromosomes are most unpacked. Resolution down to 25 kb is possible, but chromosome morphology is lost so there are no external reference points against which to map the position of the probe. This technique is therefore used after preliminary map information has been obtained, usually as a means of determining the order of a series of markers in a small region of a chromosome.

Interphase chromosomes contain the most unpacked of all cellular DNA molecules. To improve the resolution of FISH to better than 25 kb it is therefore necessary to abandon intact chromosomes and instead use purified DNA. This approach, called fiber-FISH, makes use of DNA prepared by gel stretching or molecular combing and can distinguish markers that are less than 10 kb apart.

Sequence Tagged Site (STS) Mapping

To generate a detailed physical map of a large genome we need, ideally, a high-resolution mapping procedure that is rapid and not technically demanding. Neither of the two techniques that we have considered so far - restriction mapping and FISH - meets these requirements. Restriction mapping is rapid, easy, and provides detailed information, but it cannot be applied to large genomes. FISH can be applied to large genomes, and modified versions such as fiber-FISH can give high-resolution data, but FISH is difficult to carry out and data accumulation is slow, map positions for no more than three or four markers being obtained in a single experiment. If detailed physical maps are to become a reality then we need a more powerful technique.

At present the most powerful physical mapping technique, and the one that has been responsible for generation of the most detailed maps of large genomes, is STS mapping. A sequence tagged site or STS is simply a short DNA sequence, generally between 100 and 500 bp in length, that is easily recognizable and occurs only once in the chromosome or genome being studied. To map a set of STSs, a collection of overlapping DNA fragments from a single chromosome or from the entire genome is needed. A fragment collection has been prepared from a single chromosome, with each point along the chromosome represented on average five times in the collection.

The data from which the map will be derived are obtained by determining which fragments contain which STSs. This can be done by hybridization analysis but PCR is generally used because it is quicker and has proven to be more amenable to automation. The chances of two STSs being present on the same fragment will, of course, depend on how close together they are in the genome. If they are very close then there is a good chance that they will always be on the same fragment; if they are further apart then sometimes they will be on the same fragment and sometimes they will not. The data can therefore be used to calculate the distance between two markers, in a manner analogous to the way in which map distances are determined by linkage analysis (Section 5.2.3). Remember that in linkage analysis a map distance is calculated from the frequency at which crossovers occur between two markers. STS mapping is essentially the same, except that each map distance is based on the frequency at which breaks occur between two markers. A fragment collection suitable for STS mapping. The fragments span the entire length of a chromosome, with each point on the chromosome present in an average of five fragments. The description of STS mapping given above leaves out some critical questions: What exactly is an STS? How is the DNA fragment collection obtained?

Any Unique DNA Sequence can be Used as an STS

To qualify as an STS, a DNA sequence must satisfy two criteria. The first is that its sequence must be known, so that a PCR assay can be set up to test for the presence or absence of the STS on different DNA fragments.

The second requirement is that the STS must have a unique location in the chromosome being studied, or in the genome as a whole if the DNA fragment set covers the entire genome. If the STS sequence occurs at more than one position then the mapping data will be ambiguous. Care must therefore be taken to ensure that STSs do not include sequences found in repetitive DNA.

These are easy criteria to satisfy and STSs can be obtained in many ways, the most common sources being expressed sequence tags (ESTs), SSLPs, and random genomic sequences.

- Expressed sequence tags (ESTs). These are short sequences obtained by analysis of cDNA clones (Marra et al., 1998). Complementary DNA is prepared by converting an mRNA preparation into double-stranded DNA because the mRNA in a cell is derived from protein-coding genes, cDNAs and the ESTs obtained from them represent the genes that were being expressed in the cell from which the mRNA was prepared. ESTs are looked upon as a rapid means of gaining access to the sequences of important genes, and they are valuable even if their sequences are incomplete. An EST can also be used as an STS, assuming that it comes from a unique gene and not from a member of a gene family in which all the genes have the same or very similar sequences.
- SSLPs . In Section 5.2.2 we examined the use of microsatellites and other SSLPs in genetic mapping. SSLPs can also be used as STSs in physical mapping. SSLPs that are polymorphic and have already been mapped by linkage analysis are particularly valuable as they provide a direct connection between the genetic and physical maps.
- Random genomic sequences . These are obtained by sequencing random pieces of cloned genomic DNA, or simply by downloading sequences that have been deposited in the databases.

One method for preparing cDNA. Most eukaryotic mRNAs have a poly(A) tail at their 32 end (Section 10.1.2). This series of A nucleotides is used as the priming site for the first stage of cDNA synthesis, carried out by reverse transcriptase - a.

Fragments of DNA for STS Mapping

The second component of an STS mapping procedure is the collection of DNA fragments spanning the chromosome or genome being studied. This collection is sometimes called the mapping reagent and at present there are two ways in which it can be assembled: as a clone library and as a panel of radiation hybrids. We will consider radiation hybrids first.

A radiation hybrid is a rodent cell that contains fragments of chromosomes from a second organism (McCarthy, 1996). The technology was first developed in the 1970s when it was discovered that exposure of human cells to X-ray doses of 3000–8000 rads causes the chromosomes to break up randomly into fragments, larger X-ray doses producing smaller fragments.

This treatment is of course lethal for the human cells, but the chromosome fragments can be propagated if the irradiated cells are subsequently fused with non-irradiated hamster or other rodent cells. Fusion is stimulated either chemically with polyethylene glycol or by exposure to Sendai virus. Not all of the hamster cells take up chromosome fragments so a means of identifying the hybrids is needed. The routine selection process is to use a hamster cell line that is unable to make either thymidine kinase (TK) or hypoxanthine phosphoribosyl transferase (HPRT), deficiencies in either of these two enzymes being lethal when the cells are grown in a medium containing a mixture of hypoxanthine, aminopterin and thymidine (HAT medium). After fusion, the cells are placed in HAT medium.

Those that grow are hybrid hamster cells that have acquired human DNA fragments that include genes for the human TK and HPRT enzymes, which are synthesized inside the hybrids, enabling these cells to grow in the selective medium. The treatment results in hybrid cells that contain a random selection of human DNA fragments inserted into the hamster chromosomes.

Typically the fragments are 5–10 Mb in size, with each cell containing fragments equivalent to 15–35% of the human genome. The collection of cells is called a radiation hybrid panel and can be used as a mapping reagent in STS mapping, provided that the PCR assay used to identify the STS does not amplify the equivalent region of DNA from the hamster genome.

The radiation hybrid map of the mouse genome. Physical mapping is a prelude to sequencing of the mouse genome and enables comparisons to be made between mouse and human chromosomes.

A second type of radiation hybrid panel, containing DNA from just one human chromosome, can be constructed if the cell line that is irradiated is not a human one but a second type of rodent hybrid. Cytogeneticists have developed a number of rodent cell lines in which a single human chromosome is stably propagated in the rodent nucleus. If a cell line of this type is irradiated and fused with hamster cells, then the hybrid hamster cells obtained after selection will contain either human or mouse chromosome fragments, or a mixture of both.

The ones containing human DNA can be identified by probing with a human-specific genome-wide repeat sequence, such as the short interspersed nuclear element (SINE) called Alu (Section 2.4.2), which has a copy number of just over 1 million and so occurs on average once every 4 kb in the human genome. Only cells containing human DNA will hybridize to Alu probes, enabling the uninteresting mouse hybrids to be discarded and STS mapping to be directed at the cells containing human chromosome fragments.

Radiation hybrid mapping of the human genome was initially carried out with chromosome-specific rather than whole-genome panels because it was thought that fewer hybrids would be needed to map a single chromosome than would be needed to map the entire genome. It turns out that a high-resolution map of a single human chromosome requires a panel of 100–200 hybrids, which is about the most that can be handled conveniently in a PCR screening program.

But whole-genome and single-chromosome panels are constructed differently, the former involving irradiation of just human DNA, and the latter requiring irradiation of a mouse cell containing much mouse DNA and relatively little human DNA. This means that the human DNA content per hybrid is much lower in a single-chromosome panel than in a whole-genome panel.

It transpires that detailed mapping of the entire human genome is possible with fewer than 100 whole-genome radiation hybrids, so whole-genome mapping is no more difficult than single-chromosome mapping. Once this was realized, whole-genome radiation hybrids became a central component of the mapping phase of the Human Genome Project (Section 6.3.1). Whole-genome libraries are also being used for STS mapping of other mammalian genomes and for those of the zebra fish and the chicken (McCarthy, 1996).

A clone library can also be used as the mapping reagent for STS analysis A preliminary to the sequencing phase of a genome project is to break the genome or isolated chromosomes into fragments and to clone each one in a high-capacity vector, one able to handle large fragments of DNA (Section 4.2.1). This results in a clone library, a collection of DNA fragments, which, in this case, have an average size of several hundred kb. As well as supporting the sequencing work, this type of clone library can also be used as a mapping reagent in STS analysis. As with radiation hybrid panels, a clone library can be prepared from genomic DNA, and so represents the entire genome, or a chromosome-specific library can be made if the starting DNA comes from just one type of chromosome. The latter is possible because individual chromosomes can be separated by flow cytometry. To carry out this technique, dividing cells (ones with condensed chromosomes) are carefully broken open so that a mixture of intact chromosomes is obtained. The chromosomes are then stained with a fluorescent dye.

The amount of dye that a chromosome binds depends on its size, so larger chromosomes bind more dye and fluoresce more brightly than smaller ones. The chromosome preparation is diluted and passed through a fine aperture, producing a stream of droplets, each one containing a single chromosome.

The droplets pass through a detector that measures the amount of fluorescence, and hence identifies which droplets contain the particular chromosome being sought.

An electric charge is applied to these drops, and no others, enabling the droplets containing the desired chromosome to be deflected and separated from the rest. What if two different chromosomes have similar sizes, as is the case with human chromosomes 21 and 22? These can usually be separated if the dye that is used is not one that binds non-specifically to DNA, but instead has a preference for AT- or GC-rich regions. Examples of such dyes are Hoechst 33258 and chromomycin A_3, respectively.

Two chromosomes that are the same size rarely have identical GC contents, and so can be distinguished by the amounts of AT- or GC-specific dye that they bind. Separating chromosomes by flow cytometry. A mixture of fluorescently stained chromosomes is passed through a small aperture so that each drop that emerges contains just one chromosome. The fluorescence detector identifies the signal from drops containing.

Compared with radiation hybrid panels, clone libraries have one important advantage for STS mapping. This is the fact that the individual clones can subsequently provide the DNA that is actually sequenced. The data resulting from STS analysis, from which the physical map is generated, can equally well be used to determine which clones contain overlapping DNA fragments, enabling a clone contig to be built up. This assembly of overlapping clones can be used as the base material for a lengthy, continuous DNA sequence, and the STS data can later be used to anchor this sequence precisely onto the physical map. If the STSs also include SSLPs that have been mapped by genetic linkage analysis then the DNA sequence, physical map and genetic map can all be integrated.

The value of clone libraries in genome projects. The small clone library shown in this example contains sufficient information for an STS map to be constructed, and can also be used as the source of the DNA that will be sequenced.

Gene Interaction.

A gene interaction is an interplay between multiple genes that has an impact on the expression of an organism's phenotype. While the expression of physical traits is often described as the result of inheriting two genes, one at each allele from each parent, it is actually much more complicated.

Groups of genes interact with each other, explaining why phenotypes are so variable between individual members of a species. Understanding gene interactions is an important aspect of understanding inheritance, particularly inheritance of deleterious traits.

Gene interactions can result in the alteration or suppression of a phenotype.

This can occur when an organism inherits two different dominant genes, for example, resulting in incomplete dominance. This is commonly seen in flowers, where breeding two flowers that pass down dominant genes can result in a flower of an unusual color caused by incomplete dominance.

If red and white are dominant, for example, the offspring might be pinkish or striped in color as the result of a gene interaction. Sometimes, genetic traits are entirely suppressed. People with albinism may carry genes for traits that are not expressed in their phenotypes because the albinism acts to turn those genes off.

This is also seen in coloration patterns in animals such as tortoiseshell cats, where the unusual hair color is the result of selective gene interactions, with genes being turned off at some locations and turned on in others.

The fruit fly is famously extensively studied in genetics and much of the understanding of how gene interactions works comes from working with the fruit fly in lab environments. In organisms like humans, where genetic experimentation is viewed as unethical, geneticists are forced to rely on data from the existing population to learn about dominant and recessive traits and to see how groups of genetic traits can interact.

A gene interaction is the result of inheriting genes that conflict in some way, making it impossible for all of them to express as coded, or of inheriting a set of interrelated genes that interact with each other to express a phenotype.

Sometimes a gene interaction limits production of certain proteins, often quite early in fetal development. In other instances, it can interfere with the coding of proteins to result in a garbled expression of a physical trait.

Hybrids often show a variety of interesting results of gene interactions. In some cases, the interactions are beneficial and may develop into their own genetic traits, while in others, the gene interaction may create a disadvantage and those traits will eventually die out.

6

Structural Bioinformatic Approaches

Prediction of Protein Structure

Protein structure prediction is another important application of bioinformatics. The amino acid sequence of a protein, the so-called primary structure, can be easily determined from the sequence on the gene that codes for it. In the vast majority of cases, this primary structure uniquely determines a structure in its native environment. (Of course, there are exceptions, such as the bovine spongiform encephalopathy – a.k.a.

Mad Cow Disease – prion.) Knowledge of this structure is vital in understanding the function of the protein. For lack of better terms, structural information is usually classified as one of secondary, tertiary and quaternary structure. A viable general solution to such predictions remains an open problem.

As of now, most efforts have been directed towards heuristics that work most of the time. One of the key ideas in bioinformatics is the notion of homology. In the genomic branch of bioinformatics, homology is used to predict the function of a gene: if the sequence of gene A, whose function is known, is homologous to the sequence of gene B, whose function is unknown, one could infer that B may share A's function. In the structural branch of bioinformatics, homology is used to determine which parts of a protein are important in structure formation and interaction with other proteins. In a technique called homology modeling, this information is used to predict the structure of a protein once the structure of a homologous protein is known. This currently remains the only way to predict protein structures reliably.

One example of this is the similar protein homology between hemoglobin in humans and the hemoglobin in legumes (leghemoglobin). Both serve the same purpose of transporting oxygen in the organism. Though both of these proteins have completely different amino acid sequences, their protein structures are virtually identical, which reflects their near identical purposes.

Other techniques for predicting protein structure include protein threading and de novo (from scratch) physics-based modeling.

Molecular Interaction

Efficient software is available today for studying interactions among proteins, ligands and peptides. Types of interactions most often encountered in the field include – Protein–ligand (including drug), protein–protein and protein–peptide. Molecular dynamic simulation of movement of atoms about rotatable bonds is the fundamental principle behind computational algorithms, termed docking algorithms for studying molecular interactions.

Docking algorithms

In the last two decades, tens of thousands of protein three-dimensional structures have been determined by X-ray crystallography and Protein nuclear magnetic resonance spectroscopy (protein NMR). One central question for the biological scientist is whether it is practical to predict possible protein–protein interactions only based on these 3D shapes, without doing protein–protein interaction experiments. A variety of methods have been developed to tackle the Protein–protein docking problem, though it seems that there is still much work to be done in this field.

Software and Tools

Software tools for bioinformatics range from simple command-line tools, to more complex graphical programs and standalone web-services available from various bioinformatics companies or public institutions.

Open Source Bioinformatics Software

Many free and open source software tools have existed and continued to grow since the 1980s. The combination of a continued need for new algorithms for the analysis of emerging types of biological readouts, the potential for innovative in silico experiments, and freely available open code bases have helped to create opportunities for all research groups to contribute to both bioinformatics and the range of open source software available, regardless of their funding arrangements. The open source tools often act as incubators of ideas, or community-supported plug-ins in commercial applications. They may also provide de facto standards and shared object models for assisting with the challenge of bioinformation integration.

The range of open source software packages includes titles such as Bioconductor, BioPerl, BioJava, BioRuby, Bioclipse, EMBOSS, Taverna workbench, and UGENE. In order to maintain this tradition and create further opportunities, the non-profit Open Bioinformatics Foundation have supported the annual Bioinformatics Open Source Conference (BOSC) since 2000.

Web Services in Bioinformatics

SOAP and REST-based interfaces have been developed for a wide variety of bioinformatics applications allowing an application running on one computer in one part of the world to use algorithms, data and computing resources on servers in other parts of the world. The main advantages derive from the fact that end users do not have to deal with software and database maintenance overheads. Basic bioinformatics services are classified by the EBI into three categories: SSS (Sequence Search Services), MSA (Multiple Sequence Alignment) and BSA (Biological Sequence Analysis). The availability of these service-oriented bioinformatics resources demonstrate the applicability of web based bioinformatics solutions, and range from a collection of standalone tools with a common data format under a single, standalone or web-based interface, to integrative, distributed and extensible bioinformatics workflow management systems.

Proteogenomics

Proteogenomics is an emerging field of biological research at the intersection of proteomics and genomics. While this intersection is large and can be defined in multiple ways, the term proteogenomics commonly refers to studies that use proteomic information, often derived from mass spectrometry, to improve gene annotations.

Practical Applications

Proteogenomics has been applied to improve the gene annotations of various organisms. The term proteogenomics was first used in this context by a Harvard team in 2004, although the research in this field had been building up in the previous decade. Since then, the approach has been extended to other species including Arabidopsis thaliana, humans, multiple species of Shewanella bacteria, chicken, among many others. Besides improving gene annotations, proteogenomic studies can also provide valuable information about the presence of programmed frameshifts, N-terminal methionine excision, signal peptides, proteolysis and other posttranslational modifications.

Methodology

The main idea behind the proteogenomic approach is to identify peptides in a biological sample using mass spectrometry by searching the six-frame translation of the genome sequence, as opposed to searching the protein database.

This enables identification of protein regions that are absent from or incorrectly represented in current gene annotations, and thus allows improvement of the gene annotations. Comparative proteogenomics is a branch of proteogenomics that compares proteomic data from multiple related species concurrently and exploits the homology between their proteins to improve annotations with higher statistical confidence.

Immunoproteomics

Immunoproteomics is a term used to describe the study of large sets of proteins (proteomics) involved in the immune response.

Examples of common applications of immunoproteomics include:

- the isolation and mass spectrometric identification of MHC (major histocompatibility complex) binding peptides
- purification and identification of protein antigens binding specific antibodies (or other affinity reagents), and
- comparative immunoproteomics to identify proteins and pathways modulated by a specific infectious organism, disease or toxin.

The term proteomics also usually implies that mass spectrometry is the ultimate technique used for protein identification.

Metabolomics

Metabolomics is the scientific study of chemical processes involving metabolites. Specifically, metabolomics is the "systematic study of the unique chemical fingerprints that specific cellular processes leave behind", the study of their small-molecule metabolite profiles. The metabolome represents the collection of all metabolites in a biological cell, tissue, organ or organism, which are the end products of cellular processes.

Thus, while mRNA gene expression data and proteomic analyses do not tell the whole story of what might be happening in a cell, metabolic profiling can give an instantaneous snapshot of the physiology of that cell. One of the challenges of systems biology and functional genomics is to integrate proteomic, transcriptomic, and metabolomic information to give a more complete picture of living organisms.

Origins

The idea that biological fluids reflect the health of an individual has existed for a long time. Ancient Chinese doctors used ants for the evaluation of urine of patients to detect whether the urine contained high levels of glucose, and hence detect diabetes. In the Middle Ages, "urine charts" were used to link the colours, tastes and smells of urine to various medical conditions, which are metabolic in origin.

The concept that individuals might have a "metabolic profile" that could be reflected in the makeup of their biological fluids was introduced by Roger Williams in the late 1940s, who used paper chromatography to suggest characteristic metabolic patterns in urine and saliva were associated with diseases such as schizophrenia. However, it was only through technological advancements in the 1960s and 1970s that it became feasible to quantitatively (as opposed to qualitatively) measure metabolic profiles. The term "metabolic profile" was introduced by Horning, et al. in 1971 after they demonstrated that gas chromatography-mass spectrometry (GC-MS) could be used to measure compounds present in human urine and tissue extracts. The Horning group, along with that of Linus Pauling and Arthur Robinson led the development of GC-MS methods to monitor the metabolites present in urine through the 1970s.

Concurrently, NMR spectroscopy, which was discovered in the 1940s, was also undergoing rapid advances. In 1974, Seeley et al. demonstrated the utility of using NMR to detect metabolites in unmodified biological samples.

This first study on muscle highlighted the value of NMR in that it was determined that 90% of cellular ATP is complexed with magnesium. As sensitivity has improved with the evolution of higher magnetic field strengths and magic-angle spinning, NMR continues to be a leading analytical tool to investigate metabolism. Recent efforts to utilize NMR for metabolomics have been largely driven by the laboratory of Dr. Jeremy Nicholson at Birkbeck College, University of London and later at Imperial College London. In 1984, Nicholson showed H NMR spectroscopy could potentially be used to diagnose and treat diabetes mellitus, and later pioneered the application of pattern recognition methods to NMR spectroscopic data.

In 2005, the first metabolomics web database, METLIN, for characterizing human metabolites was developed in the Siuzdak laboratory at The Scripps Research Institute and contained over 5,000 metabolites and tandem mass spectral data. As of 2011, METLIN contains over 40,000 metabolites as well as the largest repository of tandem mass spectrometry data in metabolomics.

On 23 January 2007, the Human Metabolome Project, led by Dr. David Wishart of the University of Alberta, Canada, completed the first draft of the human metabolome, consisting of a database of approximately 2500 metabolites, 1200 drugs and 3500 food components. Similar projects have been underway in several plant species, most notably Medicago truncatula and Arabidopsis thaliana for several years. As late as mid-2010, metabolomics was still considered an "emerging field". Further, it was noted that further progress in the field depended in large part, through addressing otherwise "irresolvable technical challenges", by technical evolution of mass spectrometry instrumentation.

Metabolome

Metabolome refers to the complete set of small-molecule metabolites (such as metabolic intermediates, hormones and other signaling molecules, and secondary metabolites) to be found within a biological sample, such as a single organism. The word was coined in analogy with transcriptomics and proteomics; like the transcriptome and the proteome, the metabolome is dynamic, changing from second to second. Although the metabolome can be defined readily enough, it is not currently possible to analyse the entire range of metabolites by a single analytical method. The first metabolite database(called METLIN) for searching m/z values from mass spectrometry data was developed by scientists at The Scripps Research Institute in 2005. In January 2007, scientists at the University of Alberta and the University of Calgary completed the first draft of the human metabolome. They catalogued approximately 2500 metabolites, 1200 drugs and 3500 food components that can be found in the human body, as reported in the literature. This information, available at the Human Metabolome Database (www.hmdb.ca) and based on analysis of information available in the current scientific literature, is far from complete. In contrast, much more is known about the metabolomes of other organisms. For example, over 50,000 metabolites have been characterized from the plant kingdom, and many thousands of metabolites have been identified and/or characterized from single plants.

Metabolites

Metabolites are the intermediates and products of metabolism. Within the context of metabolomics, a metabolite is usually defined as any molecule less than 1 kDa in size. However, there are exceptions to this depending on the sample and detection method. For example, macromolecules such as lipoproteins and albumin are reliably detected in NMR-based metabolomics studies of blood plasma. In plant-based metabolomics, it is common to refer to "primary" and "secondary" metabolites. A primary metabolite is directly involved in the normal growth, development, and reproduction. A secondary metabolite is not directly involved in those processes, but usually has important ecological function. Examples include antibiotics and pigments. By contrast, in human-based metabolomics, it is more common to describe metabolites as being either endogenous (produced by the host organism) or exogenous. Metabolites of foreign substances such as drugs are termed xenometabolites. The metabolome forms a large network of metabolic reactions, where outputs from one enzymatic chemical reaction are inputs to other chemical reactions. Such systems have been described as hypercycles.

Metabonomics

Metabonomics is defined as "the quantitative measurement of the dynamic multiparametric metabolic response of living systems to pathophysiological stimuli or genetic modification". The word origin is from the Greek meta meaning change and nomos meaning a rule set or set of laws. This approach was pioneered by Jeremy Nicholson at Imperial College London and has been used in toxicology, disease diagnosis and a number of other fields. Historically, the metabonomics approach was one of the first methods to apply the scope of systems biology to studies of metabolism.

There has been some disagreement over the exact differences between 'metabolomics' and 'metabonomics'. The difference between the two terms is not related to choice of analytical platform: although metabonomics is more associated with NMR spectroscopy and metabolomics with mass spectrometry-based techniques, this is simply because of usages amongst different groups that have popularized the different terms. While there is still no absolute agreement, there is a growing consensus that 'metabolomics' places a greater emphasis on metabolic profiling at a cellular or organ level and is primarily concerned with normal endogenous metabolism. 'Metabonomics' extends metabolic profiling to include information about perturbations of metabolism caused by environmental factors (including diet and toxins), disease processes, and the involvement of extragenomic influences, such as gut microflora. This is not a trivial difference; metabolomic studies should, by definition, exclude metabolic contributions from extragenomic sources, because these are external to the system being studied. However, in practice, within the field of human disease research there is still a large degree of overlap in the way both terms are used, and they are often in effect synonymous.

Analytical Technologies

Separation Methods

- Gas chromatography, especially when interfaced with mass spectrometry (GC-MS), is one of the most widely used and powerful methods. It offers very high chromatographic resolution, but requires chemical derivatization for many biomolecules: only volatile chemicals can be analysed without derivatization. (Some modern instruments allow '2D' chromatography, using a short polar column after the main analytical column, which increases the resolution still further.) Some large and polar metabolites cannot be analysed by GC.
- High performance liquid chromatography (HPLC). Compared to GC, HPLC has lower chromatographic resolution, but it does have the advantage that a much wider range of analytes can potentially be measured.
- Capillary electrophoresis (CE). CE has a higher theoretical separation efficiency than HPLC, and is suitable for use with a wider range of metabolite classes than is GC. As for all electrophoretic techniques, it is most appropriate for charged analytes.

Detection Methods

- Mass spectrometry (MS) is used to identify and to quantify metabolites after separation by GC, HPLC (LC-MS), or CE. GC-MS is the most 'natural' combination of the three, and was the first to be developed. In addition, mass spectral fingerprint libraries exist or can be developed that allow identification of a metabolite according to its fragmentation pattern. MS is both sensitive (although, particularly for HPLC-MS, sensitivity is more of an issue as it is affected by the charge on the metabolite, and can be subject to ion suppression artifacts) and can be very specific. There are also a number of studies which use MS as a stand-alone technology: the sample is infused directly into the mass spectrometer with no prior separation, and the MS serves to both separate and to detect metabolites.
- Surface-based mass analysis has seen a resurgence in the past decade, with new MS technologies focused on increasing sensitivity, minimizing background, and reducing sample preparation. The ability to analyze metabolites directly from biofluids and tissues continues to challenge current MS technology, largely because of the limits imposed by the complexity of these samples, which contain thousands to tens of thousands of metabolites. Among the technologies being developed to address this challenge is Nanostructure-Initiator MS (NIMS), a desorption/ ionization approach that does not require the application of matrix and thereby facilitates small-molecule (i.e., metabolite) identification. MALDI is also used however, the application of a MALDI matrix can add significant background at <1000 Da that complicates analysis of the low-mass range (i.e., metabolites). In addition, the size of the resulting matrix crystals limits the spatial resolution that can be achieved in tissue imaging. Because of these limitations, several other matrix-free desorption/ionization approaches have been applied to the analysis of biofluids and tissues. Secondary ion mass spectrometry (SIMS) was one of the first matrix-free desorption/ionization approaches used to analyze metabolites from biological samples. SIMS uses a high-energy primary ion beam to desorb and generate secondary ions from a surface. The primary advantage of SIMS is its high spatial resolution (as small as 50 nm), a powerful characteristic for tissue imaging with MS. However, SIMS has yet to be readily applied to the analysis of biofluids and tissues because of its limited sensitivity at >500 Da and analyte fragmentation generated by the high-energy primary ion beam. Desorption electrospray ionization (DESI) is a matrix-free technique for analyzing biological samples that uses a charged solvent spray to

desorb ions from a surface. Advantages of DESI are that no special surface is required and the analysis is performed at ambient pressure with full access to the sample during acquisition. A limitation of DESI is spatial resolution because "focusing" the charged solvent spray is difficult. However, a recent development termed laser ablation ESI (LAESI) is a promising approach to circumvent this limitation.

- Nuclear magnetic resonance (NMR) spectroscopy. NMR is the only detection technique which does not rely on separation of the analytes, and the sample can thus be recovered for further analyses. All kinds of small molecule metabolites can be measured simultaneously - in this sense, NMR is close to being a universal detector. The main advantages of NMR are high analytical reproducibility and simplicity of sample preparation. Practically, however, it is relatively insensitive compared to mass spectrometry-based techniques.
- Although NMR and MS are the most widely used techniques, other methods of detection that have been used include ion-mobility spectrometry, electrochemical detection (coupled to HPLC) and radiolabel (when combined with thin-layer chromatography).

Statistical Methods

The data generated in metabolomics usually consist of measurements performed on subjects under various conditions. These measurements may be digitized spectra, or a list of metabolite levels. In its simplest form this generates a matrix with rows corresponding to subjects and columns corresponding to metabolite levels. Several statistical programs are currently available for analysis of both NMR and mass spectrometry data. For mass spectrometry data, software is available that identifies molecules that vary in subject groups on the basis of mass and sometimes retention time depending on the experimental design. The first comprehensive software to analyze global mass spectrometry-based metabolomics datasets was developed by the Siuzdak laboratory at The Scripps Research Institute in 2006.

This software, called XCMS, is freely available, has over 20,000 downloads since its inception in 2006, and is one of the most widely cited mass spectrometry-based metabolomics software programs in scientific literature. Other popular metabolomics programs for mass spectral analysis are MZmine, MetAlign, MathDAMP, which also compensate for retention time deviation during sample analysis. LCMStats is another R package for detailed analysis of liquid chromatography mass spectrometry(LCMS)data and is helpful in identification of co-eluting ions especially isotopologues from a complicated metabolic profile. It combines xcms package functions and can be used to apply many statistical functions for correcting detector saturation using coates correction and creating heat plots. Metabolomics data may also be analyzed by statistical projection (chemometrics) methods such as principal components analysis and partial least squares regression.

Key Applications

- Toxicity assessment/toxicology. Metabolic profiling (especially of urine or blood plasma samples) can be used to detect the physiological changes caused by toxic insult of a chemical (or mixture of chemicals). In many cases, the observed changes can be related to specific syndromes, e.g. a specific lesion in liver or kidney. This is of particular relevance to pharmaceutical companies wanting to test the toxicity of potential drug candidates: if a compound can be eliminated before it reaches clinical trials on the grounds of adverse toxicity, it saves the enormous expense of the trials.
- Functional genomics. Metabolomics can be an excellent tool for determining the phenotype caused by a genetic manipulation, such as gene deletion or insertion. Sometimes this can be a sufficient goal in itself—for instance, to detect any phenotypic changes in a genetically-modified plant intended for human or animal consumption. More exciting is the prospect of predicting the function of unknown genes by comparison with the metabolic perturbations caused by deletion/insertion of known genes.

  Such advances are most likely to come from model organisms such as Saccharomyces cerevisiae and Arabidopsis thaliana. The Cravatt laboratory at The Scripps Research Institute has recently applied this technology to mammalian systems, identifying the N-acyltaurines as previously uncharacterized endogenous substrates for the enzyme fatty acid amide hydrolase (FAAH) and the monoalkylglycerol ethers (MAGEs) as endogenous substrates for the uncharacterized hydrolase KIAA1363.
- Nutrigenomics is a generalised term which links genomics, transcriptomics, proteomics and metabolomics to human nutrition. In general a metabolome in a given body fluid is influenced by endogenous factors such as age, sex, body composition and genetics as well as underlying pathologies. The large bowel microflora are also a very significant potential confounder of metabolic profiles and could be classified as either an endogenous or exogenous factor. The main exogenous factors are diet and drugs. Diet can then be broken down to nutrients and non- nutrients. Metabolomics is one means to determine a biological endpoint, or metabolic fingerprint, which reflects the balance of all these forces on an individual's metabolism.

Environmental Metabolomics

- Environmental Metabolomics is the application of metabolomics to characterise the interactions of organisms with their environment. This approach has many advantages for studying organism–environment interactions and for assessing organism function and health at the molecular level. As such, metabolomics is finding an increasing number of applications in the environmental sciences, ranging from understanding organismal responses to abiotic pressures, to investigating the responses of organisms to other biota. These interactions can be studied from individuals to populations, which can be related to the traditional fields of ecophysiology and ecology, and from instantaneous effects to those over evolutionary time scales, the latter enabling studies of genetic adaptation.

Lipidomics

Lipidomics may be defined as the large-scale study of pathways and networks of cellular lipids in biological systems The word "lipidome" is used to describe the complete lipid profile within a cell, tissue or organism and is a subset of the "metabolome" which also includes the three other major classes of biological molecules: proteins/amino-acids, sugars and nucleic acids. Lipidomics is a relatively recent research field that has been driven by rapid advances in technologies such as mass spectrometry (MS), nuclear magnetic resonance (NMR) spectroscopy, fluorescence spectroscopy, dual polarisation interferometry and computational methods, coupled with the recognition of the role of lipids in many metabolic diseases such as obesity, atherosclerosis, stroke, hypertension and diabetes. This rapidly expanding field complements the huge progress made in genomics and proteomics, all of which constitute the family of systems biology. Lipidomics research involves the identification and quantification of the thousands of cellular lipid molecular species and their interactions with other lipids, proteins, and other metabolites. Investigators in lipidomics examine the structures, functions, interactions, and dynamics of cellular lipids and the changes that occur during perturbation of the system.

Han and Gross first defined the field of lipidomics through integrating the specific chemical properties inherent in lipid molecular species with a comprehensive mass spectrometric approach. Although lipidomics is under the umbrella of the more general field of "metabolomics", lipidomics is itself a distinct discipline due to the uniqueness and functional specificity of lipids relative to other metabolites. In lipidomic research, a vast amount of information quantitatively describing the spatial and temporal alterations in the content and composition of different lipid molecular species is accrued after perturbation of a cell through changes in its physiological or pathological state. Information obtained from these studies facilitates mechanistic insights into changes in cellular function.

Therefore, lipidomic studies play an essential role in defining the biochemical mechanisms of lipid-related disease processes through identifying alterations in cellular lipid metabolism, trafficking and homeostasis. The growing attention on lipid research is also seen from the initiatives underway of the LIPID Metabolites And Pathways Strategy (LIPID MAPS Consortium). and The European Lipidomics Initiative (ELIfe).

Structural Diversity of Lipids

Lipids are a diverse and ubiquitous group of compounds which have many key biological functions, such as acting as structural components of cell membranes, serving as energy storage sources and participating in signaling pathways. Lipids may be broadly defined as hydrophobic or amphipathic small molecules that originate entirely or in part from two distinct types of biochemical subunits or "building blocks": ketoacyl and isoprene groups.

The huge structural diversity found in lipids arises from the biosynthesis of various combinations of these building blocks. For example, glycerophospholipids are composed of a glycerol backbone linked to one of approximately 10 possible headgroups and also to 2 fatty acyl/alkyl chains, which in turn may have 30 or more different molecular structures. In practice, not all possible permutations are detected experimentally, due to chain preferences depending on the cell type and also to detection limits - nevertheless several hundred distinct glycerophospholipid molecular species have been detected in mammalian cells

Experimental Techniques

Lipid Extraction

Most methods of lipid extraction and isolation from biological samples exploit the high solubility of hydrocarbon chains in organic solvents. Given the diversity in lipid classes, it is not possible to accommodate all classes with a common extraction method. The traditional Bligh/Dyer procedure uses chloroform/methanol-based protocols that include phase partitioning into the organic layer. These protocols work relatively well for a wide variety of physiologically relevant lipids but they have to be adapted for complex lipid chemistries and low-abundance and labile lipid metabolites When organic soil was used ,citrate buffer in the extraction mixture gave higher amounts of lipid phosphate than acetate buffer,Tris,H2O or phosphate buffer.

Lipid Separation

The simplest method of lipid separation is the use of thin layer chromatography (TLC). Although not as sensitive as other methods of lipid detection, it offers a rapid and comprehensive screening tool prior to more sensitive and sophisticated techniques.

Solid-phase extraction (SPE) chromatography is useful for rapid, preparative separation of crude lipid mixtures into different lipid classes. This involves the use of prepacked columns containing silica or other stationary phases to separate glycerophospholipids, fatty acids, cholesteryl esters, glycerolipids, and sterols from crude lipid mixtures. High performance liquid chromatography (HPLC or LC) is extensively used in lipidomic analysis to separate lipids prior to mass analysis. Separation can be achieved by either normal-phase HPLC or reverse-phase HPLC. For example, normal phase HPLC effectively separates glycerophospholipids on the basis of headgroup polarity, whereas reverse-phase HPLC effectively separates fatty acids such as eicosanoids on the basis of chain length, degree of unsaturation and substitution. HPLC of lipids may either be performed offline or online where the eluate is integrated with the ionization source of a mass spectrometer.

Lipid Detection

The progress of modern lipidomics has been greatly accelerated by the development of spectrometric methods in general and soft ionization techniques for mass spectrometry such as electrospray ionization (ESI) and matrix-assisted laser desorption/ionization (MALDI) in particular. "Soft" ionization does not cause extensive fragmentation, so that comprehensive detection of an entire range of lipids within a complex mixture can be correlated to experimental conditions or disease state. In addition to ESI and MALDI, the technique of atmospheric pressure chemical ionization (APCI) has become increasingly popular for the analysis of nonpolar lipids.

Detection of a fatty acid by LC-MS/MS

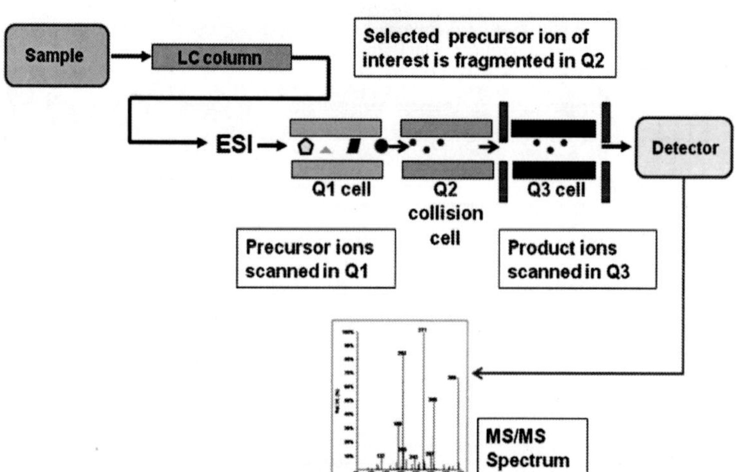

Figure: *Schema showing detection of a fatty acid by LC-MS/MS using a linear ion-trap instrument and an electrospray (ESI) ion source.*

ESI MS

ESI-MS was initially developed by Fenn and colleagues for analysis of biomolecules. It depends on the formation of gaseous ions from polar, thermally labile and mostly non-volatile molecules and thus is completely suitable for a variety of lipids. It is a soft-ionization method that rarely disrupts the chemical nature of the analyte prior to mass analysis. Various ESI-MS methods have been developed for analysis of different classes, subclasses, and individual lipid species from biological extracts. Comprehensive reviews of the methods and their application have recently been published. The major advantages of ESI-MS are high accuracy, sensitivity, reproducibility, and the applicability of the technique to complex solutions without prior derivatization. Han and coworkers have developed a method known as"shotgun lipidomics" which involves direct infusion of a crude lipid extract into an ESI source optimized for intrasource separation of lipids based on their intrinsic electrical properties.

MALDI MS

MALDI mass spectrometry is a laser-based soft-ionization method often used for analysis of large proteins, but has been used successfully for lipids. The lipid is mixed with a matrix, such as 2,5-dihydroxybenzoic acid, and applied to a sample holder as a small spot. A laser is fired at the spot, and the matrix absorbs the energy, which is then transferred to the analyte, resulting in ionization of the molecule. MALDI-Time-of-flight (MALDI-TOF) MS has become a very promising approach for lipidomics studies, particularly for the imaging of lipids from tissue slides.

APCI MS

The source for APCI is similar to ESI except that ions are formed by the interaction of the heated analyte solvent with a corona discharge needle set at a high electrical potential. Primary ions are formed immediately surrounding the needle, and these interact with the solvent to form secondary ions that ultimately ionize the sample. APCI is particularly useful for the analysis of nonpolar lipids such as triacylglycerols, sterols, and fatty acid esters.

Imaging Techniques

Recent developments in MALDI methods have enabled direct detection of lipids in-situ. Abundant lipid-related ions are produced from the direct analysis of thin tissue slices when sequential spectra are acquired across a tissue surface that has been coated with a MALDI matrix. Collisional activation of the molecular ions can be used to determine the lipid family and often structurally define the molecular species. This technique enables detection of phospholipids, sphingolipids and glycerolipids in tissues such as heart, kidney and brain.

Lipidomic Profiling

Lipid profiling is a targeted metabolomics platform that provides a comprehensive analysis of lipid species within a cell or tissue. Profiling based on electrospray ionization tandem mass spectrometry (ESI-MS/MS) is capable of providing quantitative data and is adaptable to high throughput analyses.

The powerful approach of transgenics, namely deletion and/or overexpression of a gene product coupled with lipidomics, can give valuable insights into the role of biochemical pathways. Lipid profiling techniques have also been applied to plants and microorganisms such as yeast. A combination of quantitative lipidomic data in conjunction with the corresponding transcriptional data (using gene-array methods) and proteomic data (using tandem MS) enables a systems biology approach to a more in-depth understanding of the metabolic or signaling pathways of interest.

Informatics

A major challenge for lipidomics, in particular for MS-based approaches, lies in the computational and bioinformatic demands of handling the large amount of data that arise at various stages along the chain of information acquisition and processing. Chromatographic and MS data collection requires substantial efforts in spectral alignment and statistical evaluation of fluctuations in signal intensities. Such variations have a multitude of origins, including biological variations, sample handling and analytical accuracy. As a consequence several replicates are normally required for reliable determination of lipid levels in complex mixtures. Within the last few years, a number of software packages have been developed by various companies and research groups to analyze data generated by MS profiling of metabolites, including lipids.

The data processing for differential profiling usually proceed through several stages, including input file manipulation, spectral filtering, peak detection, chromatographic alignment, normalization, visualization, and data export. An example of metabolic profiling software is the freely-available Java-based Mzmine application. Some software packages such as Markerview include multivariate statistical analysis (for example, principal component analysis) and these will be helpful for the identification of correlations in lipid metabolites that are associated with a physiological phenotype, in particular for the development of lipid-based biomarkers.Another objective of the information technology side of lipidomics involves the construction of metabolic maps from data on lipid structures and lipid-related protein and genes. Some of these lipid pathways are extremely complex, for example the mammalian glycosphingolipid pathway. The establishment of searchable and interactive databases of lipids and lipid-related genes/proteins is also an extremely important resource as a reference for the lipidomics community. Integration of these databases with

MS and other experimental data, as well as with metabolic networks offers an opportunity to devise therapeutic strategies to prevent or reverse these pathological states involving dysfunction of lipid-related processes.

Shotgun Proteomics

Shotgun proteomics is a method of identifying proteins in complex mixtures using a combination of high performance liquid chromatography combined with mass spectrometry. The name is derived from shotgun sequencing of DNA which is itself named by analogy with the rapidly-expanding, quasi-random firing pattern of a shotgun. In shotgun proteomics, the proteins in the mixture are digested and the resulting peptides are separated by liquid chromatography. Tandem mass spectrometry is then used to identify the peptides.

Top-down Proteomics

Top-down proteomics is a method of protein identification that uses an ion trapping mass spectrometer to store an isolated protein ion for mass measurement and tandem mass spectrometry analysis. The name is derived from the similar approach to DNA seqencing. Proteins are typically ionized by electrospray ionization and trapped in a Fourier transform ion cyclotron resonance (Penning trap) or quadrupole ion trap (Paul trap) mass spectrometer. Fragmentation for tandem mass spectrometry is accomplished by electron-capture dissociation or electron-transfer dissociation. Recently, a top-down approach has been developed for mapping the connectivity of the disulfide-rich peptide hedyotide B2 . This method allows rapid characterization of the disulfide pattern of cystine-knot miniproteins such as cyclotides, conotoxin, knottin and plant defensins.

Bottom-up Proteomics

Bottom-up proteomics is a common method to identify proteins and characterize their amino acid sequences and post-translational modifications by proteolytic digestion of proteins prior to analysis by mass spectrometry. The proteins may first be purified by a method such as gel electrophoresis resulting in one or a few proteins in each proteolytic digest. Alternatively, the crude protein extract is digested directly, followed by one or more dimensions of separation of the peptides by liquid chromatography coupled to mass spectrometry, a technique known as shotgun proteomics. By comparing the masses of the proteolytic peptides or their tandem mass spectra with those predicted from a sequence database or annotated peptide spectral in a peptide spectral library, peptides can be identified and multiple peptide identifications assembled into a protein identification.

Transcriptomics

The transcriptome is the set of all RNA molecules, including mRNA, rRNA, tRNA, and other non-coding RNA produced in one or a population of cells.

Scope

The term can be applied to the total set of transcripts in a given organism, or to the specific subset of transcripts present in a particular cell type. Unlike the genome, which is roughly fixed for a given cell line (excluding mutations), the transcriptome can vary with external environmental conditions. Because it includes all mRNA transcripts in the cell, the transcriptome reflects the genes that are being actively expressed at any given time, with the exception of mRNA degradation phenomena such as transcriptional attenuation. The study of transcriptomics, also referred to as expression profiling, examines the expression level of mRNAs in a given cell population, often using high-throughput techniques based on DNA microarray technology. The use of next-generation sequencing technology to study the transcriptome at the nucleotide level is known as RNA-Seq.

Analysis

DNA microarrays can provide a method for comparing on a genome-wide basis the abundance of DNAs in the same samples. The DNA in spots can only be PCR products that are specific for individual genes. A DNA copy of RNA is made using the enzyme reverse transcriptase. Gene arrays can only be used to investigate genes in eukaryotic cells. Sequencing is now being used instead of gene arrays to quantify DNA levels, at least semi quantitatively.

A number of organism-specific transcriptome databases have been constructed and annotated to aid in the identification of genes that are differentially expressed in distinct cell populations or subtypes.

Applications

The transcriptomes of stem cells and cancer cells are of particular interest to researchers who seek to understand the processes of cellular differentiation and carcinogenesis. Analysis of the transcriptomes of human oocytes and embryos is used to understand the molecular mechanisms and signaling pathways controlling early embryonic development, and could theoretically be a powerful tool in making proper embryo selection in in vitro fertilisation.

Relation to Proteome

Proteome

The transcriptome can be seen as a precursor for the proteome, that is, the entire set of proteins expressed by a genome. However, the analysis of relative mRNA expression levels can be complicated by the fact that relatively small changes in mRNA expression can produce large changes in the total amount of the corresponding protein present in the cell. One analysis method, known as Gene Set Enrichment Analysis, identifies coregulated gene networks rather than individual genes that are up- or down-regulated in different cell populations. Although microarray studies can reveal the relative amounts of different mRNAs in the cell, levels of mRNA are not directly proportional to the expression level of the proteins they code for. The number of protein molecules synthesized using a given mRNA molecule as a template is highly dependent on translation-initiation features of the mRNA sequence; in particular, the ability of the translation initiation sequence is a key determinant in the recruiting of ribosomes for protein translation. The complete protein complement of a cell or organism is known as the proteome. A study of 158,807 mouse transcripts revealed that 4,520 of these transcripts form antisense partners that are base pair complementary to the exons of genes. These results raise the possibility that significant numbers of "antisense RNA-coding genes" might participate in the regulation of the levels of expression of protein-coding mRNAs.

Phosphoproteomics

Phosphoproteomics is a branch of proteomics that identifies, catalogs, and characterizes proteins containing a phosphate group as a post-translational modification. Phosphorylation is a key reversible modification that regulates protein function, subcellular localization, complex formation, degradation of proteins and therefore cell signalling networks. With all of these modification results, it is assumed that up to 30% of all proteins may be phosphorylated, some multiple times. Compared to expression analysis, phosphoproteomics provides two additional layers of information. First, it provides clues on what protein or pathway might be activated because a change in phosphorylation status almost always reflects a change in protein activity. Second, it indicates what proteins might be potential drug targets as exemplified by the kinase inhibitor Gleevec. While phosphoproteomics will greatly expand knowledge about the numbers and types of phosphoproteins, its greatest promise is the rapid analysis of entire phosphorylation based signalling networks.

Overview of Phosphoproteomic Analysis

A sample large-scale phosphoproteomic analysis includes

1. Cultured cells undergo SILAC encoding.
2. Cells are stimulated with factor of interest (eg. growth factor, hormone).
3. Stimulation can occur for various lengths of time for temporal analysis.
4. Cells are lysed and enzymatically digested.
5. Peptides are separated using ion exchange chromatography.
6. Phosphopeptides are enriched using phosphospecific antibodies, immobilized metal affinity chromatography or titanium dioxide (TiO_2) chromatography.
7. Phosphopeptides are analyzed using mass spectrometry.
8. Peptides are sequenced and analyzed.

Tools and Methods

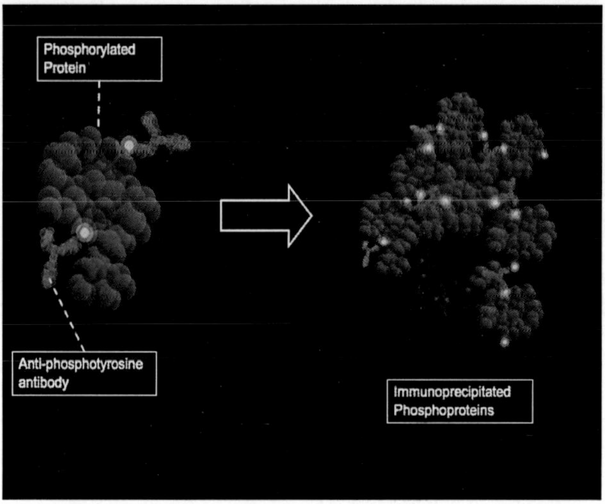

Figure: *Method of phosphoprotein purification by immunoprecipitation with anti-phososhotyrosine antibodies.*

The analysis of the entire complement of phosphorylated proteins in a cell is certainly a feasible option. This is due to the optimization of enrichment protocols for phosphoproteins and phosphopeptides, better fractionation techniques using chromatography, and improvement of methods to selectively visualize phosphorylated residues using mass spectrometry. Although the current procedures for phosphoproteomic analysis are greatly improved, there is still sample loss and inconsistencies with regards to sample preparation, enrichment , and instrumentation. Bioinformatics tools and biological sequence databases are also necessary for high-throughput phosphoproteomic studies.

Enrichment Strategies

Previous procedures to isolate phosphorylated proteins included radioactive labeling with P-labeled ATP followed by SDS polyacrylamide gel electrophoresis or thin layer chromatography. These traditional methods are inefficient because it is impossible to obtain large amounts of proteins required for phosphorylation analysis. Therefore, the current and simplest methods to enrich phosphoproteins are affinity purification using phosphospecific antibodies, immobilized metal affinity chromatography (IMAC), strong cation exchange (SCX) chromatography, or titanium dioxide chromatography. Antiphosphotyrosine antibodies have been proven very successful in purification, but fewer reports have been published using antibodies against phosphoserine- or phosphothreonine-containing proteins. IMAC enrichment is based on phosphate affinity for immobilized metal chelated to the resin. SCX separates phosphorylated from non-phosphorylated peptides based on the negatively charged phosphate group. Titanium dioxide chromatography is a newer technique that requires significantly less column preparation time. Many phosphoproteomic studies use a combination of these enrichment strategies to obtain the purest sample possible.

Mass Spectrometry Analysis

Mass spectrometry is currently the best method to adequately compare pairs of protein samples. The two main procedures to perform this task are using isotope-coded affinity tags (ICAT) and stable isotopic amino acids in cell culture (SILAC). In the ICAT procedure samples are labeled individually after isolation with mass-coded reagents that modify cysteine residues. In SILAC, cells are cultured separately in the presence of different isotopically labeled amino acids for several cell divisions allowing cellular proteins to incorporate the label. Mass spectrometry is subsequently used to identify phosphoserine, phosphothreonine, and phosphotyrosine-containing peptides.

Phosphoproteomics in the Study of Signal Transduction

Intracellular signal transduction is primarily mediated by the reversible phosphorylation of various signalling molecules by enzymes dubbed kinases. Kinases transfer phosphate groups from ATP to specific serine, threonine or tyrosine residues of target molecules. The resultant phosphorylated protein may have altered activity level, subcellular localization or tertiary structure.

Phosphoproteomic analyses are ideal for the study of the dynamics of signalling networks. In one study design, cells are exposed to SILAC labelling and then stimulated by a specific growth factor. The cells are collected at various timepoints, and the lysates are combined for analysis by tandem MS. This allows experimenters to track the phosphorylation state of many phosphoproteins in the cell over time. The ability to measure the global phosphorylation state of many proteins at various time points makes this approach much more powerful than traditional biochemical methods for analyzing signalling network behavior.

One study was able to simultaneously measure the fold-change in phosphorylation state of 127 proteins between unstimulated and EphrinB1-stimulated cells. Of these 127 proteins, 40 showed increased phosphorylation with stimulation by EphrinB1. The researchers were able to use this information in combination with previously published data to construct a signal transduction network for the proteins downstream of the EphB2 receptor.

Another recent phosphoproteomic study included large-scale identification and quantification of phosphorylation events triggered by the anti-diuretic hormone vasopressin in kidney collecting duct. A total of 714 phosphorylation sites on 223 unique phosphoproteins were identified, including three novel phosphorylation sites in the vasopressin-sensitive water channel aquaporin-2 (AQP2).

Phosphoproteomics in the Study of Cancer

Since the inception of phosphoproteomics, cancer research has focused on changes to the phosphoproteome during tumor development. Phosphoproteins could be cancer markers useful to cancer diagnostics and therapeutics. In fact, research has shown that there are distinct phosphotyrosine proteomes of breast and liver tumors. There is also evidence of hyperphosphorylation at tyrosine residues in breast tumors but not in normal tissues. Findings like these suggest that it is possible to mine the tumor phosphoproteome for potential biomarkers.

Increasing amounts of data are available suggesting that distinctive phosphoproteins exist in various tumors and that phosphorylation profiling could be used to fingerprint cancers from different origins. In addition, systematic cataloguing of tumor-specific phosphoproteins in individual patients could reveal multiple causative players during cancer formation. By correlating this experimental data to clinical data such as drug response and disease outcome, potential cancer markers could be identified for diagnosis, prognosis, prediction of drug response, and potential drug targets.

Limitations

While phosphoproteomics has greatly expanded knowledge about the numbers and types of phosphoproteins, along with their role in signaling networks, there are still a few limitations to these techniques. To begin with, isolation methods such as anti-phosphotyrosine antibodies do not distinguish between isolating tyrosine-phosphorylated proteins and proteins associated with tyrosine-phosphorylated proteins. Therefore, even though phosphorylation dependent protein-protein interactions are very important, it is important to remember that a protein detected by this method is not necessarily a direct substrate of any tyrosine kinase. Only by digesting the samples before immunoprecipitation can isolation of only phosphoproteins and temporal profiles of individual phosphorylaiton sites be produced.

Another limitation is that some relevant proteins will likely be missed since no extraction condition is all encompassing. It is possible that proteins with low stoichiometry of phosphorylation, in very low abundance, or phosphorylated as a target for rapid degradation will be lost.

PEGylation

PEGylation is the process of covalent attachment of polyethylene glycol polymer chains to another molecule, normally a drug or therapeutic protein. PEGylation is routinely achieved by incubation of a reactive derivative of PEG with the target macromolecule. The covalent attachment of PEG to a drug or therapeutic protein can "mask" the agent from the host's immune system (reduced immunogenicity and antigenicity), and increase the hydrodynamic size (size in solution) of the agent which prolongs its circulatory time by reducing renal clearance. PEGylation can also provide water solubility to hydrophobic drugs and proteins.

History

In 1970s, pioneering research by Dr. Frank Davis, Dr. Abraham Abuchowski and colleagues foresaw the potential of the conjugation of Polyethylene glycol (PEG) to proteins. Dr. Abuchowski founded Enzon, Inc, which brought three PEGylated drugs to market. He is the founder and president of Prolong Pharmaceuticals.

PEGylation, is a process of attaching the strands of the polymer PEG to molecules most typically peptides, proteins, and antibody fragments, that can help to meet the challenges of improving the safety and efficiency of many therapeutics.

It produces alterations in the physiochemical properties including changes in conformation, electrostatic binding, hydrophobicity etc. These physical and chemical changes increase systemic retention of the therapeutic agent. Also, it can influence the binding affinity of the therapeutic moiety to the cell receptors and can alter the absorption and distribution patterns.

PEGylation, by increasing the molecular weight of a molecule, can impart several significant pharmacological advantages over the unmodified form, such as:

- Improved drug solubility
- Reduced dosage frequency, without diminished efficacy with potentially reduced toxicity
- Extended circulating life
- Increased drug stability
- Enhanced protection from proteolytic degradation

PEGylated drugs have the following commercial advantages also:

- Opportunities for new delivery formats and dosing regimens
- Extended patent life of previously approved drugs

PEGylated Pharmaceuticals on the Market

The clinical value of PEGylation is now well established. ADAGEN (PEG-bovine adenosine deaminase) manufactured by Enzon Pharmaceuticals, Inc., US was the first PEGylated protein approved by the U.S. Food and Drug Administration (FDA) in March 1990, to enter the market. It is used to treat X-linked severe combined immunogenicity syndrome, as an alternative to bone marrow transplantation and enzyme replacement by gene therapy. Since the introduction of ADAGEN, a large number of PEGylated protein and peptide pharmaceuticals have followed and many others are under clinical trial or under development stages. Some of the successful examples are:

- PEGASYS: PEGylated interferon alpha for use in the treatment of chronic hepatitis C and hepatitis B (Hoffmann-La Roche)
- Pegintron: PEGylated interferon alpha for use in the treatment of chronic hepatitis C and hepatitis B (Schering-Plough / Enzon)
- Oncaspar: PEGylated L-asparaginase for the treatment of acute lymphoblastic leukemia in patients who are hypersensitive to the native unmodified form of L-asparaginase (Enzon). This drug was recently approved for front line use.
- Neulasta: PEGylated recombinant methionyl human granulocyte colony-stimulating factor for severe cancer chemotherapy-induced neutropenia (Amgen)
- Doxil/Caelyx: PEGylated liposome containing doxorubicin for the treatment of cancer (Ortho Biotech / Schering-Plough)
- Pegloticase (Krystexxa) - PEGylated Uricase for the treatment of gout (Savient)

PEG Moiety Properties

PEG is a particularly attractive polymer for conjugation. The specific characteristics of PEG moieties relevant to pharmaceutical applications are:

- Water solubility
- High mobility in solution
- Lack of toxicity and immunogenicity
- Ready clearance from the body
- Altered distribution in the body

PEGylation Process

The first step of the PEGylation is the suitable functionalization of the PEG polymer at one or both terminals. PEGs that are activated at each terminus with the same reactive moiety are known as "homobifunctional", whereas if the functional groups present are different, then the PEG derivative is referred as "heterobifunctional" or "heterofunctional." The chemically active or activated derivatives of the PEG polymer are prepared to attach the PEG to the desired molecule. The overall PEGylation processes used to date for protein conjugation can be broadly classified into two types, namely a solution phase batch process and an on-column fed-batch process. The simple and commonly adopted batch process involves the mixing of reagents together in a suitable buffer solution, preferably at a temperature between 4 and 6°C, followed by the separation and purification of the desired product using a suitable technique based on its physicochemical properties, including size exclusion chromatography (SEC), ion exchange chromatography (IEX), hydrophobic interaction chromatography (HIC) and membranes or aqueous two phase systems.

The choice of the suitable functional group for the PEG derivative is based on the type of available reactive group on the molecule that will be coupled to the PEG. For proteins, typical reactive amino acids include lysine, cysteine, histidine, arginine, aspartic acid, glutamic acid, serine, threonine, tyrosine. The N-terminal amino group and the C-terminal carboxylic acid can also be used as a site specific site by conjugation with aldehyde functional polymers.

The techniques used to form first generation PEG derivatives are generally reacting the PEG polymer with a group that is reactive with hydroxyl groups, typically anhydrides, acid chlorides, chloroformates and carbonates. In the second generation PEGylation chemistry more efficient functional groups such as aldehyde, esters, amides etc made available for conjugation.

As applications of PEGylation have become more and more advanced and sophisticated, there has been an increase in need for heterobifunctional PEGs for conjugation. These heterobifunctional PEGs are very useful in linking two entities, where a hydrophilic, flexible and biocompatible spacer is needed. Preferred end groups for heterobifunctional PEGs are maleimide, vinyl sulfones, pyridyl disulfide, amine, carboxylic acids and NHS esters.

Third generation pegylation agents, where the shape of the polymer has been branched, Y shaped or comb shaped are available which show reduced viscosity and lack of organ accumulation.

Future Perspectives

Even though nearly four decades of development in PEGylation technology has proven its pharmacological advantages and acceptability, the technology still lags in providing a commercially attractive, generic process to produce highly specific PEGylated therapeutic products at high yield. As a multi-million dollar annual business with growing interest from both emerging biotechnology and established multinational pharmaceutical companies, there is great scientific and commercial interest in improving present methodologies and in introducing innovative process variations.

Activity Based Proteomics

Activity based proteomics, or activity based protein profiling (ABPP) is a functional proteomic technology that uses specially designed chemical probes that react with mechanistically-related classes of enzymes. The basic unit of

ABPP is the probe which typically consists of two elements: a reactive group (RG) and a tag. Additionally, some probes may contain a binding group which enhances selectivity. The reactive group usually contains an electrophile that gets covalently-linked to a nucleophilic residue in the active site of an active enzyme. An enzyme that is inhibited by enzyme inhibitors or post-translational modifications will not react with an activity-based probe. The tag may be either a reporter such as a fluorophore or an affinity label such as biotin or an alkyne or azide for use with the Huisgen 1,3-dipolar cycloaddition (also known as click chemistry). A major advantage of ABPP is the ability to monitor the availability of the enzyme active site directly, rather than being limited to protein or mRNA abundance. With classes of enzymes such as the serine proteases and metalloproteases that often interact with endogenous inhibitors or that exist as inactive zymogens, this technique offers a valuable advantage over traditional techniques that rely on abundance rather than activity.

Finally, in recent years ABPP has been combined with tandem mass spectrometry enabling the identification of hundreds of active enzymes from a single sample. This technique, known as ABPP-MudPIT is especially useful for profiling inhibitor selectivity as the potency of an inhibitor can be tested against hundreds of targets simultaneously. ABPP was originally invented by James C. Powers (Georgia Tech, Atlanta) in the early 1990s and subsequently adopted and developed by the other labs, for example Benjamin Cravatt III at The Scripps Research Institute, Matthew Bogyo at Stanford, and others.

Protein Databases

UniProt

UniProt is a comprehensive, high-quality and freely accessible database of protein sequence and functional information, many of which are derived from genome sequencing projects. It contains a large amount of information about the biological function of proteins derived from the research literature.

The UniProt Consortium

The UniProt Consortium comprises the European Bioinformatics Institute (EBI), the Swiss Institute of Bioinformatics (SIB), and the Protein Information Resource (PIR). EBI, located at the Wellcome Trust Genome Campus in Hinxton, UK, hosts a large resource of bioinformatics databases and services. SIB, located in Geneva, Switzerland, maintains the ExPASy (Expert Protein Analysis System) servers that are a central resource for proteomics tools and databases.

PIR, hosted by the National Biomedical Research Foundation (NBRF) at the Georgetown University Medical Center in Washington, DC, USA, is heir to the oldest protein sequence database, Margaret Dayhoff's Atlas of Protein Sequence and Structure, first published in 1965. In 2002, EBI, SIB, and PIR joined forces as the UniProt Consortium.

The Roots of UniProt Databases

Each consortium member is heavily involved in protein database maintenance and annotation. Until recently, EBI and SIB together produced the Swiss-Prot and TrEMBL databases, while PIR produced the Protein Sequence Database (PIR-PSD). These databases coexisted with differing protein sequence coverage and annotation priorities.

Swiss-Prot was created in 1986 by Amos Bairoch during his PhD and developed by the Swiss Institute of Bioinformatics and subsequently developed by Rolf Apweiler at the European Bioinformatics Institute. Swiss-Prot aimed to provide reliable protein sequences associated with a high level of annotation (such as the description of the function of a protein, its domain structure, post-translational modifications, variants, etc.), a minimal level of redundancy and high level of integration with other databases.

Recognizing that sequence data were being generated at a pace exceeding Swiss-Prot's ability to keep up, TrEMBL (Translated EMBL Nucleotide Sequence Data Library) was created to provide automated annotations for those proteins not in Swiss-Prot. Meanwhile, PIR maintained the PIR-PSD and related databases, including iProClass, a database of protein sequences and curated families. The consortium members pooled their overlapping resources and expertise, and launched UniProt in December 2003.

Organization of UniProt Databases

UniProt provides four core databases:

UniProtKB

UniProt Knowledgebase (UniProtKB) is a protein database curated by experts, consisting of two sections. UniProtKB/Swiss-Prot (containing reviewed, manually annotated entries) and UniProtKB/TrEMBL (containing unreviewed, automatically annotated entries). In release 2010_09 of 10 August 2010, UniProtKB/Swiss-Prot contained 519,348 entries, and UniProtKB/TrEMBL contained 11,636,205 entries.

UniProtKB/Swiss-Prot

UniProtKB/Swiss-Prot is a high-quality, manually annotated, non-redundant protein sequence database. It combines information extracted from scientific literature and biocurator-evaluated computational analysis. The aim of UniProtKB/Swiss-Prot is to provide all known relevant information about a particular protein. Annotation is regularly reviewed to keep up with current scientific findings. The manual annotation of an entry involves detailed analysis of the protein sequence and of the scientific literature.

Sequences from the same gene and the same species are merged into the same database entry. Differences between sequences are identified, and their cause documented (for example alternative splicing, natural variation, incorrect initiation sites, incorrect exon boundaries, frameshifts, unidentified conflicts). A range of sequence analysis tools is used in the annotation of UniProtKB/Swiss-Prot entries. Computer-predictions are manually evaluated, and relevant results selected for inclusion in the entry. These predictions include post-translational modifications, transmembrane domains and topology, signal peptides, domain identification, and protein family classification. Relevant publications are identified by searching databases such as PubMed. The full text of each paper is read, and information is extracted and added to the entry. Annotation arising from the scientific literature includes, but is not limited to:

- Protein and gene names
- Function
- Enzyme-specific information such as catalytic activity, cofactors and catalytic residues
- Subcellular location
- Protein-protein interactions
- Pattern of expression
- Locations and roles of significant domains and sites
- Ion-, substrate- and cofactor-binding sites
- Protein variant forms produced by natural genetic variation, RNA editing, alternative splicing, proteolytic processing, and post-translational modification

Annotated entries undergo quality assurance before inclusion into UniProtKB/Swiss-Prot. When new data becomes available, entries are updated.

UniProtKB/TrEMBL

UniProtKB/TrEMBL contains high-quality computationally analyzed records, which are enriched with automatic annotation. It was introduced in response to increased dataflow resulting from genome projects, as the time- and labour-consuming manual annotation process of UniProtKB/Swiss-Prot could not be broadened to include all available protein sequences. The translations of annotated coding sequences in the EMBL-Bank/GenBank/DDBJ nucleotide sequence database are automatically processed and entered in UniProtKB/TrEMBL. UniProtKB/TrEMBL also contains sequences from PDB, and from gene prediction, including Ensembl, RefSeq and CCDS.

UniParc

UniProt Archive (UniParc) is a comprehensive and non-redundant database, which contains all the protein sequences from the main, publicly available protein sequence databases. Proteins may exist in several different source databases, and in multiple copies in the same database. In order to avoid redundancy, UniParc stores each unique sequence only once. Identical sequences are merged, regardless of whether they are from the same or different species. Each sequence is given a stable and unique identifier (UPI), making it possible to identify the same protein from different source databases. UniParc contains only protein sequences, with no annotation. When sequences in the source databases change, these changes are tracked by UniParc and history of all changes is archived.

Source Databases

Currently UniParc contains protein sequences from the following publicly available databases:

- INSDC EMBL-Bank/DDBJ/GenBank nucleotide sequence databases
- Ensembl
- European Patent Office (EPO)
- FlyBase
- H-Invitational Database (H-Inv)
- International Protein Index (IPI)
- Japan Patent Office (JPO)
- Protein Information Resource (PIR-PSD)
- Protein Data Bank (PDB)
- Protein Research Foundation (PRF)
- RefSeq
- Saccharomyces Genome Database (SGD)
- The Arabidopsis Information Resource (TAIR)
- TROME
- US Patent Office (USPTO)
- UniProtKB/Swiss-Prot, UniProtKB/Swiss-Prot protein isoforms, UniProtKB/TrEMBL
- Vertebrate and Genome Annotation Database (VEGA)
- WormBase

UniRef

The UniProt Reference Clusters (UniRef) consist of three databases of clustered sets of protein sequences from UniProtKB and selected UniParc records. The UniRef100 database combines identical sequences and sequence fragments (from any organism) into a single UniRef entry. The sequence of a representative protein, the accession numbers of all the merged entries and links to the corresponding UniProtKB and UniParc records are displayed.

UniRef100 sequences are clustered using the CD-HIT algorithm to build UniRef90 and UniRef50. Each cluster is composed of sequences that have at least 90% or 50% sequence identity, respectively, to the longest sequence. Clustering sequences significantly reduces database size, enabling faster sequence searches.

UniMes

The UniProt Metagenomic and Environmental Sequences (UniMES) database is a repository specifically developed for metagenomic and environmental data. The predicted proteins from this dataset are combined with automatic classification by InterPro to enhance the original information with further analysis. UniProtKB contains protein sequences from known species, data arising from metagenomics studies is from environmental (i.e. uncultured) samples and as such the species may not be known/identified.

UniMES was developed for this data. Data from UniMES is not included in UniProtKB or UniRef, but is included in UniParc. UniMES includes data from the Global Ocean Sampling Expedition (GOS). UniMES is available from the UniProt FTP site

Funding for UniProt

UniProt is funded by grants from the National Human Genome Research Institute, the National Institutes of Health (NIH), the European Commission, the Swiss Federal Government through the Federal Office of Education and Science, NCI-caBIG, and the Department of Defense.

- Protein Information Resource (PIR)
- Swiss-Prot
- Protein Data Bank (PDB)
- National Center for Biotechnology Information (NCBI)

The National Center for Biotechnology Information (NCBI) is part of the United States National Library of Medicine (NLM), a branch of the National Institutes of Health. The NCBI is located in Bethesda, Maryland(38.994994°N 77.099339°WCoordinates: 38.994994°N 77.099339°W) and was founded in 1988 through legislation sponsored by Senator Claude Pepper. The NCBI houses genome sequencing data in GenBank and an index of biomedical research articles in PubMed Central and PubMed, as well as other information relevant to biotechnology. All these databases are available online through the Entrez search engine.

NCBI is directed by David Lipman, one of the original authors of the BLAST sequence alignment program and a widely respected figure in Bioinformatics.

He also leads an intramural research program, including groups led by Stephen Altschul (another BLAST co-author), David Landsman, and Eugene Koonin (a prolific author on comparative genomics).

GenBank

The NCBI has had responsibility for making available the GenBank DNA sequence database since 1992. GenBank coordinates with individual laboratories and other sequence databases such as those of the European Molecular Biology Laboratory (EMBL) and the DNA Data Bank of Japan (DDBJ).

Since 1992, NCBI has grown to provide other databases in addition to GenBank. NCBI provides Online Mendelian Inheritance in Man, the Molecular Modeling Database (3D protein structures), dbSNP a database of single-nucleotide polymorphisms, the Unique Human Gene Sequence Collection, a Gene Map of the human genome, a Taxonomy Browser, and coordinates with the National Cancer Institute to provide the Cancer Genome Anatomy Project. The NCBI assigns a unique identifier (Taxonomy ID number) to each species of organism. The NCBI has software tools that are available by WWW browsing or by FTP. For example, BLAST is a sequence similarity searching program. BLAST can do sequence comparisons against the GenBank DNA database in less than 15 seconds. The GenBank sequence database is an open access, annotated collection of all publicly available nucleotide sequences and their protein translations. This database is produced and maintained by the National Center for Biotechnology Information (NCBI) as part of the International Nucleotide Sequence Database Collaboration (INSDC). The National Center for Biotechnology Information is a part of the National Institutes of Health in the United States.

GenBank and its collaborators receive sequences produced in laboratories throughout the world from more than 100,000 distinct organisms. In more than 20 years since its establishment, GenBank has become the most important and most influential database for research in almost all biological fields, whose data were accessed and cited by millions of researchers around the world. GenBank continues to grow at an exponential rate, doubling every 18 months. Release 155, produced in August 2006, contained over 65 billion nucleotide bases in more than 61 million sequences. GenBank is built by direct submissions from individual laboratories, as well as from bulk submissions from large-scale sequencing centers.

Submissions

Only original sequences can be submitted to GenBank. Direct submissions are made to GenBank using BankIt, which is a Web-based form, or the stand-alone submission program, Sequin. Upon receipt of a sequence submission, the GenBank staff examines the originality of the data and assigns an accession number to the sequence and performs quality assurance checks. The submissions are then released to the public database, where the entries are retrievable by Entrez or downloadable by FTP. Bulk submissions of Expressed Sequence Tag (EST), Sequence-tagged site (STS), Genome Survey Sequence (GSS), and High-Throughput Genome Sequence (HTGS) data are most often submitted by large-scale sequencing centers. The GenBank direct submissions group also processes complete microbial genome sequences.

History

Walter Goad of the Theoretical Biology and Biophysics Group at Los Alamos National Laboratory and others established the Los Alamos Sequence Database in 1979, which culminated in 1982 with the creation of the public GenBank. Funding was provided by the National Institutes of Health, the National Science Foundation, the Department of Energy, and the Department of Defense. LANL collaborated on GenBank with the firm Bolt, Beranek, and Newman, and by the end of 1983 more than 2,000 sequences were stored in it.

In the mid 1980s, the Intelligenetics bioinformatics company at Stanford University managed the GenBank project in collaboration with LANL. As one of the earliest bioinformatics community projects on the Internet, the GenBank project started BIOSCI/Bionet news groups for promoting open access communications among bioscientists. During 1989 to 1992, the GenBank project transitioned to the newly created National Center for Biotechnology Information.

Growth

The GenBank release notes for release 162.0 (October 2007) state that “from 1982 to the present, the number of bases in GenBank has doubled approximately every 18 months”. The following plot clearly shows the exponential growth (on a semi-log scale such as this, a straight line represents an exponential change).

The GenBank database includes additional data sets which are constructed mechanically from the main sequence data collection, and therefore are excluded from this count.

NCBI Bookshelf

The NCBI Bookshelf is a collection of freely available, downloadable, on-line versions of selected biomedical books. As of April 2011, the Bookshelf had 845 titles covering a wide range of topics including molecular biology, biochemistry, cell biology, genetics, microbiology, disease states from a molecular and cellular point of view, research methods, and virology. Some of the books are online versions of previously published books, while others, such as Coffee Break (book), are written and edited by NCBI staff. The Bookshelf is a complement to the Entrez PubMed repository of peer-reviewed publication abstracts in that Bookshelf contents provide established perspectives on evolving areas of study and a context in which many disparate individual pieces of reported research can be organized. t

- Human Protein Reference Database
- Proteomics Identifications Database (PRIDE)
- Proteopedia The collaborative, 3D encyclopedia of proteins and other molecules
- Cardiac Organellar Protein Atlas Knowledgebase (COPaKB)

Psychogenomics/ Behavioural Genetics

Behavioural genetics is the field of study that examines the role of genetics in animal (including human) behaviour. Often associated with the "nature versus nurture" debate, behavioural genetics is highly interdisciplinary, involving contributions from biology, genetics, ethology, psychology, and statistics. Behavioural geneticists study the inheritance of behavioural traits. In humans, this information is often gathered through the use of the twin study or adoption study. In animal studies, breeding, transgenesis, and gene knockout techniques are common; psychiatric genetics is a closely related field.

History

Figure 3: *Galton in his later years*

Sir Francis Galton, a nineteenth-century intellectual, is recognized as one of the first behavioural geneticists. Galton, a cousin of Charles Darwin, studied the heritability of human ability, focusing on mental characteristics as well as eminence among close relatives in the English upper-class. In 1869, Galton published his results in Hereditary Genius. In his work, Galton "introduced multivariate analysis and paved the way towards modern Bayesian statistics" that are used throughout the sciences—launching what has been dubbed the "Statistical Enlightenment".

Behaviour genetics, per-se, gained recognition as a research discipline with the publication in 1960 of the textbook Behavior Genetics by J.L. Fuller and W.R. Thompson.

Underscoring the role of evolution in behavioural genetics, Theodosius Dobzhansky was elected the first president of the Behavior Genetics Association in 1972; the BGA bestows the Dobzhansky Award on researchers for their outstanding contributions to the field. In the early 1970s, Lee Ehrman, a doctoral student of Dobzhansky, wrote seminal papers describing the relationship between genotype frequency and mating success in Drosophila, lending impetus to the pursuit of genetic studies of behaviour in other animals.

Methods

The primary methods of modern human behavioural genetics are twin and adoption studies.

Under each of these umbrella terms, these there are large families of designs, such as discordant twin designs. These techniques have been expanded upon in recent decades to include multivariate genetic analysis, extremes analysis, and more recently genomic techniques including the Genome-wide_association_study, linkage, candidate gene approaches and others. Many of the methods of behavioural genetics are also discussed under heritability.

Notable Behavioural Geneticists

Notable behavioural geneticists include Dorret Boomsma, John DeFries, Lindon Eaves, David Fulker, John Hewitt, Kenneth Kendler, John Loehlin, Nick Martin, Gerald McClearn, Robert Plomin, Theodore Reich, who was a pioneer in psychiatric genetics, Hans van Abeelen, Avshalom Caspi, and Steven G. Vandenberg, the founding editor of the journal Behavior Genetics.

Journals

Behavioural geneticists are active in a variety of scientific disciplines including biology, medicine, pharmacology, psychiatry, and psychology; thus, behavioural-genetic research is published in a variety of scientific journals, including Nature and Science. Journals that specifically publish research in behavioural genetics include Behavior Genetics, Molecular Psychiatry, Psychiatric Genetics, Twin Research and Human Genetics, Genes, Brain and Behavior, and the Journal of Neurogenetics.

Cell

The cell is the functional basic unit of life. It was discovered by Robert Hooke and is the functional unit of all known living organisms. It is the smallest unit of life that is classified as a living thing, and is often called the building block of life.

Organisms can be classified as unicellular (consisting of a single cell; including most bacteria) or multicellular (including plants and animals). Humans contain about 100 trillion cells; a typical cell size is 10 μm and a typical cell mass is 1 nanogram.

The human cell extrema are: largest - anterior horn in the spinal cord (135 ì vs. 120 ì for the ovum), longest - pseudounipolar cells which reach from extremities, including the toes to the lower brain stem, and smallest - granule cells in the cerebellum, at 4 μ.

The largest known cells are unfertilised ostrich egg cells, which weigh 3.3 pounds. In 1835, before the final cell theory was developed, Jan Evangelista Purkynì observed small "granules" while looking at the plant tissue through a microscope.

The cell theory, first developed in 1839 by Matthias Jakob Schleiden and Theodor Schwann, states that all organisms are composed of one or more cells, that all cells come from preexisting cells, that vital functions of an organism occur within cells, and that all cells contain the hereditary information necessary for regulating cell functions and for transmitting information to the next generation of cells.

The word cell comes from the Latin cellula, meaning "a small room". The descriptive term for the smallest living biological structure was coined by Robert Hooke in a book he published in 1665 when he compared the cork cells he saw through his microscope to the small rooms monks lived in.

Anatomy

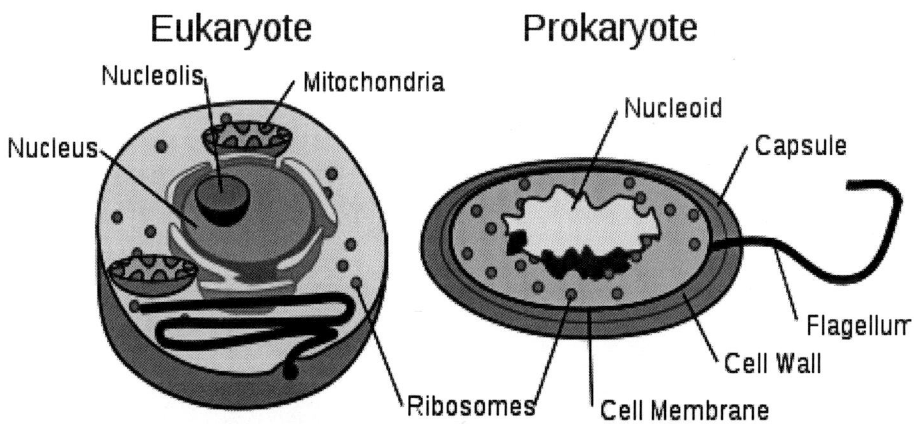

Figure: *The cells of eukaryotes (left) and prokaryotes (right)*

There are two types of cells: eukaryotic and prokaryotic. Prokaryotic cells are usually independent, while eukaryotic cells are often found in multicellular organisms.

Table: Comparison of features of prokaryotic and eukaryotic cells

	Prokaryotes	**Eukaryotes**
Typical organisms	bacteria, archaea	protists, fungi, plants, animals
Typical size	~ 1-10 µm	~ 10-100 µm (sperm cells, apart from the tail, are smaller)
Type of nucleus	nucleoid region; no real nucleus	real nucleus with double membrane
DNA	circular (usually)	linear molecules (chromosomes) with histone proteins
RNA-/protein-synthesis		coupled in cytoplasm
RNA-synthesis inside the nucleus		protein synthesis in cytoplasm
Ribosomes	50S+30S	60S+40S
Cytoplasmatic structure	very few structures	highly structured by endomembranes and a cytoskeleton
Cell movement	flagella made of flagellin	flagella and cilia containing microtubules; lamellipodia and filopodia containing actin
Mitochondria	none	one to several thousand (though some lack mitochondria)
Chloroplasts	none	in algae and plants
Organization	usually single cells	single cells, colonies, higher multicellular organisms with specialized cells
Cell division	Binary fission (simple division)	Mitosis (fission or budding) Meiosis

Prokaryotic Cells

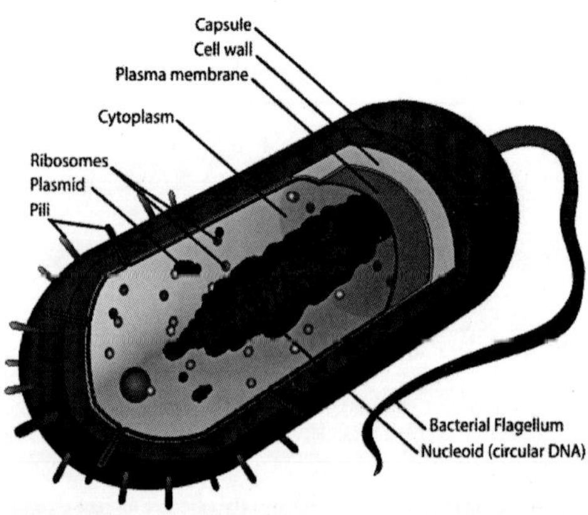

Figure: *Diagram of a typical prokaryotic cell*

The prokaryote cell is simpler, and therefore smaller, than a eukaryote cell, lacking a nucleus and most of the other organelles of eukaryotes. There are two kinds of prokaryotes: bacteria and archaea; these share a similar structure.

Nuclear material of prokaryotic cell consist of a single chromosome that is in direct contact with cytoplasm. Here, the undefined nuclear region in the cytoplasm is called nucleoid.

A prokaryotic cell has three architectural regions:

- On the outside, flagella and pili project from the cell's surface. These are structures (not present in all prokaryotes) made of proteins that facilitate movement and communication between cells;
- Enclosing the cell is the cell envelope – generally consisting of a cell wall covering a plasma membrane though some bacteria also have a further covering layer called a capsule. The envelope gives rigidity to the cell and separates the interior of the cell from its environment, serving as a protective filter.

  Though most prokaryotes have a cell wall, there are exceptions such as Mycoplasma (bacteria) and Thermoplasma (archaea). The cell wall consists of peptidoglycan in bacteria, and acts as an additional barrier against exterior forces. It also prevents the cell from expanding and finally bursting (cytolysis) from osmotic pressure against a hypotonic environment. Some eukaryote cells (plant cells and fungi cells) also have a cell wall;
- Inside the cell is the cytoplasmic region that contains the cell genome (DNA) and ribosomes and various sorts of inclusions. A prokaryotic chromosome is usually a circular molecule (an exception is that of the bacterium Borrelia burgdorferi, which causes Lyme disease). Though not forming a nucleus, the DNA is condensed in a nucleoid. Prokaryotes can carry extrachromosomal DNA elements called plasmids, which are usually circular. Plasmids enable additional functions, such as antibiotic resistance.

Eukaryotic Cells

Plants, animals, fungi, slime moulds, protozoa, & algae are all Eukaryotic. These cells are about 15 times wider than a typical prokaryote and can be as much as 1000 times greater in volume. The major difference between prokaryotes and eukaryotes is that eukaryotic cells contain membrane-bound compartments in which specific metabolic activities take place. Most important among these is a cell nucleus, a membrane-delineated compartment that houses the eukaryotic cell's DNA. This nucleus gives the eukaryote its name, which means "true nucleus." Other differences include:

- The plasma membrane resembles that of prokaryotes in function, with minor differences in the setup. Cell walls may or may not be present.
- The eukaryotic DNA is organized in one or more linear molecules, called chromosomes, which are associated with histone proteins. All chromosomal DNA is stored in the cell nucleus, separated from the cytoplasm by a membrane. Some eukaryotic organelles such as mitochondria also contain some DNA.
- Many eukaryotic cells are ciliated with primary cilia. Primary cilia play important roles in chemosensation, mechanosensation, and thermosensation. Cilia may thus be "viewed as sensory cellular antennae that coordinate a large number of cellular signaling pathways, sometimes coupling the signaling to ciliary motility or alternatively to cell division and differentiation."
- Eukaryotes can move using motile cilia or flagella. The flagella are more complex than those of prokaryotes.

Subcellular Components

All cells, whether prokaryotic or eukaryotic, have a membrane that envelops the cell, separates its interior from its environment, regulates what moves in and out (selectively permeable), and maintains the electric potential of the cell. Inside the membrane, a salty cytoplasm takes up most of the cell volume. All cells possess DNA, the hereditary material of genes, and RNA, containing the information necessary to build various proteins such as enzymes, the cell's primary machinery. There are also other kinds of biomolecules in cells. This article lists these primary components of the cell, then briefly describe their function.

Membrane

The cytoplasm of a cell is surrounded by a cell membrane or plasma membrane. The plasma membrane in plants and prokaryotes is usually covered by a cell wall. This membrane serves to separate and protect a cell from its surrounding environment and is made mostly from a double layer of lipids (hydrophobic fat-like molecules) and hydrophilic phosphorus molecules.

Hence, the layer is called a phospholipid bilayer. It may also be called a fluid mosaic membrane. Embedded within this membrane is a variety of protein molecules that act as channels and pumps that move different molecules into and out of the cell. The membrane is said to be 'semi-permeable', in that it can either let a substance (molecule or ion) pass through freely, pass through to a limited extent or not pass through at all. Cell surface membranes also contain receptor proteins that allow cells to detect external signaling molecules such as hormones.

Cytoskeleton

The cytoskeleton acts to organize and maintain the cell's shape; anchors organelles in place; helps during endocytosis, the uptake of external materials by a cell, and cytokinesis, the separation of daughter cells after cell division; and moves parts of the cell in processes of growth and mobility. The eukaryotic cytoskeleton is composed of microfilaments, intermediate filaments and microtubules. There is a great number of proteins associated with them, each controlling a cell's structure by directing, bundling, and aligning filaments. The prokaryotic cytoskeleton is less well-studied but is involved in the maintenance of cell shape, polarity and cytokinesis.

Genetic Material

Two different kinds of genetic material exist: deoxyribonucleic acid (DNA) and ribonucleic acid (RNA). Most organisms use DNA for their long-term information storage, but some viruses (e.g., retroviruses) have RNA as their genetic material. The biological information contained in an organism is encoded in its DNA or RNA sequence. RNA is also used for information transport (e.g., mRNA) and enzymatic functions (e.g., ribosomal RNA) in organisms that use DNA for the genetic code itself. Transfer RNA (tRNA) molecules are used to add amino acids during protein translation.

Prokaryotic genetic material is organized in a simple circular DNA molecule (the bacterial chromosome) in the nucleoid region of the cytoplasm. Eukaryotic genetic material is divided into different, linear molecules called chromosomes inside a discrete nucleus, usually with additional genetic material in some organelles like mitochondria and chloroplasts. A human cell has genetic material contained in the cell nucleus (the nuclear genome) and in the mitochondria (the mitochondrial genome). In humans the nuclear genome is divided into 23 pairs of linear DNA molecules called chromosomes. The mitochondrial genome is a circular DNA molecule distinct from the nuclear DNA. Although the mitochondrial DNA is very small compared to nuclear chromosomes, it codes for 13 proteins involved in mitochondrial energy production and specific tRNAs. Foreign genetic material (most commonly DNA) can also be artificially introduced into the cell by a process called transfection. This can be transient, if the DNA is not inserted into the cell's genome, or stable, if it is. Certain viruses also insert their genetic material into the genome.

Organelles

The human body contains many different organs, such as the heart, lung, and kidney, with each organ performing a different function. Cells also have a set of "little organs," called organelles, that are adapted and/or specialized for carrying out one or more vital functions. Both eukaryotic and prokaryotic cells have organelles but organelles in eukaryotes are generally more complex and may be membrane bound.

There are several types of organelles in a cell. Some (such as the nucleus and golgi apparatus) are typically solitary, while others (such as mitochondria, peroxisomes and lysosomes) can be numerous (hundreds to thousands). The cytosol is the gelatinous fluid that fills the cell and surrounds the organelles.

Cell Nucleus - Eukaryotes only - A Cell's Information Center

The cell nucleus is the most conspicuous organelle found in a eukaryotic cell. It houses the cell's chromosomes, and is the place where almost all DNA replication and RNA synthesis (transcription) occur. The nucleus is spherical and separated from the cytoplasm by a double membrane called the nuclear envelope. The nuclear envelope isolates and protects a cell's DNA from various molecules that could accidentally damage its structure or interfere with its processing. During processing, DNA is transcribed, or copied into a special RNA, called messenger RNA (mRNA). This mRNA is then transported out of the nucleus, where it is translated into a specific protein molecule. The nucleolus is a specialized region within the nucleus where ribosome subunits are assembled. In prokaryotes, DNA processing takes place in the cytoplasm.

Mitochondria and Chloroplasts - Eukaryotes only - the Power Generators

Mitochondria are self-replicating organelles that occur in various numbers, shapes, and sizes in the cytoplasm of all eukaryotic cells. Mitochondria play a critical role in generating energy in the eukaryotic cell. Mitochondria generate the cell's energy by oxidative phosphorylation, using oxygen to release energy stored in cellular nutrients (typically pertaining to glucose) to generate ATP. Mitochondria multiply by splitting in two. Respiration occurs in the cell mitochondria. Organelles that are modified chloroplasts are broadly called plastids, and are involved in energy storage through photosynthesis, which uses solar energy to generate carbohydrates and oxygen from carbon dioxide and water. Mitochondria and chloroplasts each contain their own genome, which is separate and distinct from the nuclear genome of a cell. Both organelles contain this DNA in circular plasmids, much like prokaryotic cells, strongly supporting the evolutionary theory of endosymbiosis; since these organelles contain their own genomes and have other similarities to prokaryotes, they are thought to have developed through a symbiotic relationship after being engulfed by a primitive cell.

Endoplasmic Reticulum - Eukaryotes only

The endoplasmic reticulum (ER) is the transport network for molecules targeted for certain modifications and specific destinations, as compared to molecules that float freely in the cytoplasm. The ER has two forms: the rough ER, which has ribosomes on its surface and secretes proteins into the cytoplasm, and the smooth ER, which lacks them. Smooth ER plays a role in calcium sequestration and release.

Golgi Apparatus - Eukaryotes only

The primary function of the Golgi apparatus is to process and package the macromolecules such as proteins and lipids that are synthesized by the cell. It is particularly important in the processing of proteins for secretion. The Golgi apparatus forms a part of the endomembrane system of eukaryotic cells. Vesicles that enter the Golgi apparatus are processed in a cis to trans direction, meaning they coalesce on the cis side of the apparatus and after processing pinch off on the opposite (trans) side to form a new vesicle in the animal cell.

Ribosomes

The ribosome is a large complex of RNA and protein molecules. They each consist of two subunits, and act as an assembly line where RNA from the nucleus is used to synthesise proteins from amino acids. Ribosomes can be found either floating freely or bound to a membrane (the rough endoplasmatic reticulum in eukaryotes, or the cell membrane in prokaryotes).

Lysosomes and Peroxisomes - Eukaryotes only

Lysosomes contain digestive enzymes (acid hydrolases). They digest excess or worn-out organelles, food particles, and engulfed viruses or bacteria. Peroxisomes have enzymes that rid the cell of toxic peroxides. The cell could not house these destructive enzymes if they were not contained in a membrane-bound system. These organelles are often called a "suicide bag" because of their ability to detonate and destroy the cell.

Centrosome - the Cytoskeleton Organiser

The centrosome produces the microtubules of a cell - a key component of the cytoskeleton. It directs the transport through the ER and the Golgi apparatus. Centrosomes are composed of two centrioles, which separate during cell division and help in the formation of the mitotic spindle. A single centrosome is present in the animal cells. They are also found in some fungi and algae cells.

Vacuoles

Vacuoles store food and waste. Some vacuoles store extra water. They are often described as liquid filled space and are surrounded by a membrane. Some cells, most notably Amoeba, have contractile vacuoles, which can pump water out of the cell if there is too much water. The vacuoles of eukaryotic cells are usually larger in those of plants than animals.

Structures Outside the Cell Wall

Capsule

A gelatinous capsule is present in some bacteria outside the cell wall. The capsule may be polysaccharide as in pneumococci, meningococci or polypeptide as Bacillus anthracis or hyaluronic acid as in streptococci. Capsules are not marked by ordinary stain and can be detected by special stain. The capsule is antigenic. The capsule has antiphagocytic function so it determines the virulence of many bacteria. It also plays a role in attachment of the organism to mucous membranes.

Flagella

Flagella are the organelles of cellular mobility. They arise from cytoplasm and extrude through the cell wall. They are long and thick thread-like appendages, protein in nature. Are most commonly found in bacteria cells but are found in animal cells as well.

Fimbriae (pili)

They are short and thin hair like filaments, formed of protein called pilin (antigenic). Fimbriae are responsible for attachment of bacteria to specific receptors of human cell (adherence). There are special types of pili called (sex pili) involved in conjunction.

Functions

Growth and Metabolism

Between successive cell divisions, cells grow through the functioning of cellular metabolism. Cell metabolism is the process by which individual cells process nutrient molecules.

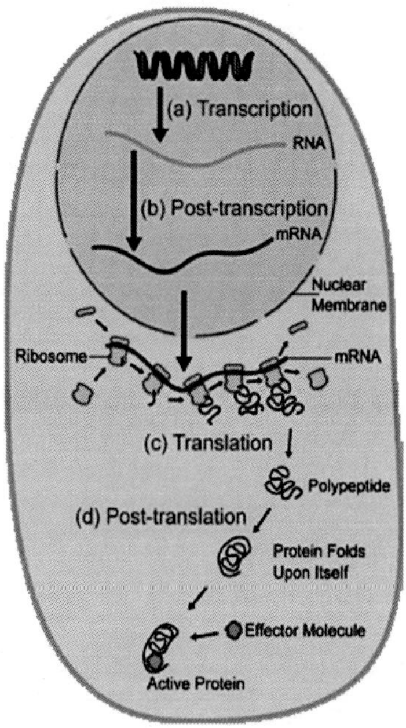

Figure: *An Overview of Protein Synthesis.*

Metabolism has two distinct divisions: catabolism, in which the cell breaks down complex molecules to produce energy and reducing power, and anabolism, in which the cell uses energy and reducing power to construct complex molecules and perform other biological functions. Complex sugars consumed by the organism can be broken down into a less chemically complex sugar molecule called glucose. Once inside the cell, glucose is broken down to make adenosine triphosphate (ATP), a form of energy, through two different pathways.

The first pathway, glycolysis, requires no oxygen and is referred to as anaerobic metabolism. Each reaction is designed to produce some hydrogen ions that can then be used to make energy packets (ATP). In prokaryotes, glycolysis is the only method used for converting energy. The second pathway, called the Krebs cycle, or citric acid cycle, occurs inside the mitochondria and can generate enough ATP to run all the cell functions.

Within the nucleus of the cell (light blue), genes (DNA, dark blue) are transcribed into RNA. This RNA is then subject to post-transcriptional modification and control, resulting in a mature mRNA (red) that is then transported out of the nucleus and into the cytoplasm (peach), where it undergoes translation into a protein. mRNA is translated by ribosomes (purple) that match the three-base codons of the mRNA to the three-base anti-codons of the appropriate tRNA. Newly synthesized proteins (black) are often further modified, such as by binding to an effector molecule (orange), to become fully active.

Creation

Cell division involves a single cell (called a mother cell) dividing into two daughter cells. This leads to growth in multicellular organisms (the growth of tissue) and to procreation (vegetative reproduction) in unicellular organisms. Prokaryotic cells divide by binary fission. Eukaryotic cells usually undergo a process of nuclear division, called mitosis, followed by division of the cell, called cytokinesis. A diploid cell may also undergo meiosis to produce haploid cells, usually four. Haploid cells serve as gametes in multicellular organisms, fusing to form new diploid cells.

DNA replication, or the process of duplicating a cell's genome, is required every time a cell divides. Replication, like all cellular activities, requires specialized proteins for carrying out the job.

Protein Synthesis

Cells are capable of synthesizing new proteins, which are essential for the modulation and maintenance of cellular activities. This process involves the formation of new protein molecules from amino acid building blocks based on information encoded in DNA/RNA. Protein synthesis generally consists of two major steps: transcription and translation. Transcription is the process where genetic information in DNA is used to produce a complementary RNA strand.

This RNA strand is then processed to give messenger RNA (mRNA), which is free to migrate through the cell. mRNA molecules bind to protein-RNA complexes called ribosomes located in the cytosol, where they are translated into polypeptide sequences. The ribosome mediates the formation of a polypeptide sequence based on the mRNA sequence. The mRNA sequence directly relates to the polypeptide sequence by binding to transfer RNA (tRNA) adapter molecules in binding pockets within the ribosome. The new polypeptide then folds into a functional three-dimensional protein molecule.

Movement or Motility

Cells can move during many processes: such as wound healing, the immune response and cancer metastasis. For wound healing to occur, white blood cells and cells that ingest bacteria move to the wound site to kill the microorganisms that cause infection.

At the same time fibroblasts (connective tissue cells) move there to remodel damaged structures. In the case of tumor development, cells from a primary tumor move away and spread to other parts of the body. Cell motility involves many receptors, crosslinking, bundling, binding, adhesion, motor and other proteins. The process is divided into three steps – protrusion of the leading edge of the cell, adhesion of the leading edge and de-adhesion at the cell body and rear, and cytoskeletal contraction to pull the cell forward. Each step is driven by physical forces generated by unique segments of the cytoskeleton.

Bibliography

Annie Tan: *Blueprints Notes and Cases - Biochemistry, Genetics and Embryology*, Blackwell Science, Delhi, 2004.

Balkwill, Fran: *DNA Is Here to Stay*, Carolrhoda Books, Inc., Minneapolis, 1993.

Banerjee, S : *Cell Biochemistry*, Dominant Pub, Delhi, 2009.

Bhatia, A L : *Biochemistry and Endocrinology*, Indus Valley, Delhi, 2002.

Brookes, Martin: *Get a Grip on Genetics, Time Life Books*, East Sussex, England, 1998.

Bruce, Donald : *Engineering Genesis: The Ethics of Genetic Engineering in Non-Human Animals*, London: Earthscan, 1998.

Cahill, Lisa: *Genetics, Theology, and Ethics: An Interdisciplinary Conversation*, New York: Crossroad, 2005.

Curtis, M.: *The Geometry of DNA: A Structural Revision*. London: Blue Gallery, 1996.

Diaz, E.: *Microbial Biodegradation: Genomics and Molecular Biology,* Caister Academic Press, UK, 2008.

Drlica, Karl : *Understanding DNA and Gene Cloning: A Guide for the Curious*, New York: Wiley, 1997.

Fukuyama F.: *Our Posthuman Future: Consequences of the Biotechnology Revolution,* Farrar, Straus and Giroux, NewYork, 2002.

Garza-Valdes, L.A.: *The DNA of God?* Hodder and Stoughton, London, 1998.

Hames, B. D.: *Genes and Embryos*, Oxford University Press, New York, 1989.

Jonoska, N.: *Self-Assembling DNA Graphs, DNA-Based Computers VIII*, Springer-Verlag, Berlin, 2003.

Kallenbach, N.R.: *Molecular Structure: Chemical Reactivity and Biological Activity,* Oxford University Press, Oxford, 1988.

Lewis, Ricki: *Human Genetics: Concepts and Applications,* Boston, Mass: McGraw Hill. 1994.

Mayr, E.: *Change of Genetic Environment and Evolution*, Allen and Unwin, London, 1954.

Nisha Khalsa: *Essentials of Biochemistry*, Aavishkar Pub, Delhi, 2008.

Old, R.W.: *Principles of Gene Manipulation*, London, Blackwell Scientific Publications, 1989.

Rabus, R: *Microbial Biodegradation: Genomics and Molecular Biology*, Caister Academic Press, UK, 2008.

Sturtevant, Alfred: *History of Genetics*, Harper and Row, New York, 1965.

Stwertka, Eve : *Genetic Engineering*, Franklin Watts, New York, 1982.

Stwertka, Eve and Albert: *Genetic Engineering*, Franklin Watts, New York, 1982.

Vasudev Bhatnagar: *Cell Science and Technology*, Campus Books, Delhi, 2009.

William, Kovacs D.: *An Introduction to Geotechnical Engineering*, Prentice Hall, UK, 1981.

Williams, J.G. and R.K. Patient : *Genetic Engineering*. Oxford, England: IRL Press, 1988.

Wilson, I.: *The Blood and The Shroud*. London: Weidenfeld & Nicolson, 1998.

Winston, R.: *The Future of Genetic Manipulation*, Phoenix, London, 1997.

Index

N

O

P

R

S

T

U

V

❑❑❑